THE
ELECTROMAGNETIC
FIELD

THE
ELECTROMAGNETIC
FIELD

N. ANDERSON, B.Sc., Ph.D.

Department of Mathematics, University of York

LONDON
LOGOS PRESS

Library of Congress
Catalog Card Number 68-16585

Published in Great Britain by
LOGOS PRESS LIMITED
in association with
ELEK BOOKS LIMITED
2 All Saints Street, London N1

Printed in Great Britain by Spottiswoode, Ballantyne and Company Limited,
London and Colchester

CONTENTS

Chapter 5

Chapter 6

Chapter 7

Chapter 8

Chapter 9

Appendix I

Appendix II

PREFACE

In this book I have tried to give an introduction to the theory of the electromagnetic field *in vacuo*. The reader is assumed to have a fairly good mathematical background. The approach I have adopted to the subject is to take the basic equations which govern the behaviour of the electromagnetic field in space and time, *in vacuo*, as postulates; I then develop and discuss the consequences of these postulates. I feel that if a student has the necessary mathematical skill, this approach enables him, in the time normally available for study of the subject, to go much further into it than he could with the more conventional approach, in which phenomena relating to static fields are discussed first and are later generalised to time-dependent fields.

Each chapter of the book contains worked examples. At the end of each chapter, there is a set of problems for the reader to attempt; for these, complete solutions are given at the end of the book. Such fully worked problems are useful to the student, and can be used to introduce some ideas not contained in the main body of the text.

I would like to thank Professor P. Hammond for giving his advice when I was writing this book, Dr. G. D. Wassermann for first really arousing my interest in the subject when I was an undergraduate, and Dr. A. M. Arthurs for reading the manuscript.

York
June 1967

N. ANDERSON

INTRODUCTION

The aim of this book is to provide a short introduction to the theory of the electromagnetic field *in vacuo*. Most books on this subject are fairly large, and students often find this a drawback, for several reasons. First, in a large book, it is often difficult, unless one is familiar with this subject, to decide which of the results given are of fundamental importance to the theory and which results are included merely to illustrate some application of the theory. Secondly, most books spend a great deal of time on the solution of boundary value problems in electrostatics and magnetostatics; while these problems will enable the reader to develop his prowess in mathematical manipulation, they add little to his knowledge of the electromagnetic field.

The method of presentation adopted is intended to keep the text concise and clear, and is as follows. It is assumed that the reader has a fair knowledge of the mathematical techniques used at this level, i.e. Fourier analysis, vector analysis and the solution of linear second order partial differential equations. The scope of the book has been limited to the electromagnetic field *in vacuo*; no discussion of the field in media is given. I believe that since, in principle, the behaviour of electromagnetic fields in matter can be derived from their behaviour *in vacuo*, it is best to have a firm understanding of the latter before considering the effect of media.

The main assumption of this book, which must be mentioned at the start, is that the equations which govern the behaviour of the electromagnetic field, that is, its development in space and time, are taken for granted. It is conventional in most treatments of the subject for these field equations to be obtained by first gradually accumulating results in electrostatics, magnetostatics and steady current flow, and then generalising to non-steady cases. This approach, while often at first appearing more acceptable to the student, has defects. The main defect is that one has to make the same, or equivalent, postulates, no matter how one proceeds. If one makes these at the outset, as here, the progress in developing their consequences is much easier and more rapid than if their introduction is delayed.

The value of a theory in physical science lies in its ability to correlate known experimental results and to predict further phenomena which can be verified by observation. The internal construction of a theory is to a large extent a matter of personal taste: the most important thing is that it should work. If postulates are made at the outset, which naturally one hopes are self-consistent, and then conclusions are derived which can be experimentally verified, then the agreement of the theoretical predictions with observation is the only justification that is needed for the postulates. It may be that there are many sets of postulates which are equivalent and which could form the basis for the theory; one then normally chooses the set which appears to be the simplest so that the calculations which will be required to extend and apply the theory are as simple as possible.

In the approach adopted here, the considerations of static fields appear as special cases of the more general treatment. This seems to me a natural presentation, although, of course, not the way in which the subject was developed originally. The relativistic four-dimensional formulation of the electromagnetic field is not considered, since students reading a first course in electromagnetic theory may not have met the special theory of relativity.

A short summary of the scope of the book may be helpful.

Chapter 1. The basic notions of charge and field intensity are considered, the equations which govern the spatial and temporal behaviour of the electromagnetic field *in vacuo* are postulated. These are the relationships between the observable quantities such as intensities, charges and currents.

Chapter 2. Some verifiable consequences of the field equations are developed and the representation of the field by potentials in order to simplify the field equations is discussed. The equations satisfied by the potentials in the most general case are derived and their solutions obtained.

Chapters 3 and 4 deal briefly with static electric and magnetic fields. The more important results are illustrated by simple applications.

Chapter 5 deals with electromagnetic waves. The understanding of these phenomena and the unification of optics and electromagnetic theory was one of the great achievements of science in the last century. The discussion here is restricted to plane harmonic waves.

Chapter 6 deals with the problem of energy and momentum residing in an electromagnetic field. Poynting's description of the flow of energy and momentum in a field is discussed, and the ideas of

Maxwell on the electromagnetic stresses which a field exerts are dealt with.

Chapter 7. The question of how a charged particle moves under various electromagnetic field configurations is considered, both from the point of view of the Newtonian equations of motion containing the Lorentz force, and from the Lagrangian formulation of analytical mechanics. Here the fields produced by the particle itself are ignored.

Chapters 8 and 9. The fields produced by a charged particle in an arbitrary state of motion are treated fully. These are the fundamental fields of classical electromagnetic theory. Attention is drawn to some of the problems of classical electromagnetic theory, problems which also occur in the present-day formulations of quantum electro-dynamics. Finally, those areas of the subject where present-day research is most concentrated are indicated.

CHAPTER 1

INTRODUCTION TO THE FIELD EQUATIONS

It is a well-established fact that bodies which are not in contact can interact with each other without the use of an intervening medium. The best known interaction of this kind is gravitation. If we describe gravitation within the framework of Newtonian mechanics, we say that a point mass A produces on a point mass B a force of attraction which acts along the line joining A to B. This force has a magnitude proportional to the ratio of the product of the masses to the square of the distance between them. This result can be extended to bodies of finite size by the well-established methods of the integral calculus.

In certain circumstances, the interaction between separated bodies can be such that the forces produced on them completely overshadow the gravitational effects. This book studies the principal interaction of this kind, the electromagnetic interaction; its aim is to formulate a theory which will enable us to describe the behaviour of any system as a result of this interaction.

1.1. CHARGE

If two bodies are capable of producing the interactions we have described, we say that they are *charged* or *magnetic*. We will learn how to distinguish the two states at a later stage. Let us for the moment consider only the case of charged bodies. It is found that a body can be charged in two different ways, or as we often say, it can have two kinds of *charge*. For instance, it may be that two charged bodies, A and B, repel each other, whereas A attracts a third charged body, C. A macroscopic body may be charged in many ways. For example, frictional forces produced by rubbing often give rise to this state; contact between a charged and an uncharged body often results in the latter becoming charged.

1

We try to describe these phenomena involving charged bodies by naming the two types of charge *positive* and *negative*. This is to be done so that like charges repel each other and unlike charges attract each other. We use the names positive and negative for the reason that charges of both kinds seem to exist in indivisible units, which are such that if we bring one unit of each kind into conjunction then they are no longer capable of interacting with another charged body and so in a way they add up to zero.

The basic units of charge are properties of the fundamental particles, which are the 'building blocks' of matter. We call the charge on a proton, a stable fundamental particle found in the nucleus of the atom and having a mass of 1.6×10^{-27} kg, the positive unit; and we call the charge on an electron, a stable fundamental particle found outside the nucleus in the atom and having a mass of 9.1×10^{-31} kg, the negative unit. These units are equal and opposite in the sense of the previous paragraph. Matter in the normal state has equal numbers of protons and electrons evenly distributed on the macroscopic scale and hence does not possess net charge; it is, as we say, *neutral*. Charge is a basic property of the fundamental particles, and one cannot 'explain' it in terms of any other known property; one merely accepts it and incorporates it into the theory.

The way in which a macroscopic body can become charged is that for some reason it may have an excess or deficiency of electrons over protons. This state may have been caused by frictional forces, as discussed earlier. Normally, the number of electrons involved in a macroscopic charging process is very large and the charge on an electron is no longer a suitable unit. We shall define a new unit of charge, the *coulomb*,* more suitable for the description of macroscopic phenomena as follows.

Imagine two equal like charged bodies (this could be deduced from the fact that they produce the same effects on a third charged body) placed at rest a distance D metres apart *in vacuo*. We shall suppose D to be very much greater than the maximum linear dimension of the charged bodies, so that we can regard them as particles. It is found experimentally that for two such bodies the force between them is inversely proportional to D^2. We then say that each body possesses q coulombs of charge if, the force exerted by one body on the other

---

* Called after C. A. Coulomb (1736–1806), French experimental physicist, noted for his meticulous work on the inverse square law in gravitation and electrostatics.

being F newtons,* we have

$$F = q^2/4\pi\epsilon_0 D^2, \tag{1.1}$$

where ϵ_0 is a constant defined by

$$\epsilon_0 = 8\cdot854 \times 10^{-12} \text{ farads}\dagger/\text{m}.$$

Thus we have defined the coulomb as a new unit of charge. It may appear at this stage as if equation (1.1) is rather a clumsy definition, but we shall see later that this choice leads to simplification of many results. The charge on a proton is $1\cdot6 \times 10^{-19}$ coulombs.

1.2. THE ELECTROMAGNETIC FIELD

Suppose that we have a system of charged bodies at rest or in motion. If we can introduce a test particle into the system, carrying a charge, say q, then we find that at any point $\mathbf{r}$ the particle experiences a force which is made up of two distinct components. The first component is independent of the motion of the particle and only depends on the position $\mathbf{r}$. The second component depends on both the position and velocity of the particle. Both components depend linearly on the magnitude of the charge on the test particle.

We find it most convenient in describing the force exerted on a charged body to define two vector fields $\mathbf{E}(\mathbf{r})$ and $\mathbf{B}(\mathbf{r})$, so that the force that would be exerted on a particle of charge q at the point $\mathbf{r}$ with velocity $\mathbf{v}$ is given by

$$\mathbf{F}(\mathbf{r}) = q[\mathbf{E}(\mathbf{r}) + \mathbf{v} \times \mathbf{B}(\mathbf{r})]. \tag{1.2}$$

This formula, due to Lorentz,‡ is of fundamental importance to the development of the theory. The vector field $\mathbf{E}(\mathbf{r})$, so defined, is called the *electric intensity*, and the field $\mathbf{B}(\mathbf{r})$ is called the *magnetic intensity* or the *magnetic induction*. We say that a region of space where at each point we have defined $\mathbf{E}(\mathbf{r})$ and $\mathbf{B}(\mathbf{r})$ is in an *electromagnetic field*.

At one time, it was thought that electromagnetic effects between charged bodies could be explained in terms of the action of a medium,

---

* Called after Sir Isaac Newton (1642–1727).
† Called after M. Faraday (1791–1867), one of the greatest experimental physicists of all time.
‡ H. A. Lorentz (1853–1928), Dutch physicist. He extended the macroscopic field theory of electromagnetism to the microscopic domain.

the so-called 'aether', which pervaded all space. This aether concept proved very misleading and has long been abandoned; we merely accept the fact that an electromagnetic field can exist *in vacuo* and that we do not need a mechanical model to account for the force (equation (1.2)) which a charged body experiences when it is in an electromagnetic field.

Equation (1.2) expresses the force on a charged particle due to the presence of other charges and defines the vector fields $\mathbf{E}(\mathbf{r})$ and $\mathbf{B}(\mathbf{r})$. However, we can look at this in a different way, which is often of great help in understanding electromagnetic phenomena. We can regard the original charge distribution as producing the physically observable fields $\mathbf{E}$ and $\mathbf{B}$; these in turn act on the charge q. This way of looking at the problem avoids the concept of action at a distance, but endows the fields with more physical reality than they previously possessed, and often seems the more acceptable point of view.

Now let us consider the units in which we measure the electric and magnetic intensities at a point. We say that an electric field at a point $\mathbf{r}$ has the intensity $\mathbf{E}(\mathbf{r})$ volts*/m if when we measure the force $\mathbf{F}(\mathbf{r})$, in newtons, exerted on a body at rest carrying q coulombs of charge, we obtain the relation:

$$\mathbf{F} = q\,\mathbf{E}\,(\mathbf{r}). \tag{1.3}$$

If a charge is moving with velocity $\mathbf{v}$, the force exerted on it at any point is given by equation (1.2).

We define the unit of magnetic induction, the weber†/m², so that the force which a field of intensity $\mathbf{B}(\mathbf{r})$ exerts on a body having a charge of q coulombs and velocity $\mathbf{v}$ m/sec at the point $\mathbf{r}$ is $q[\mathbf{v}(\mathbf{r}) \times \mathbf{B}(\mathbf{r})]$ newtons. The unit weber/m² seems an awkward unit for magnetic intensity, but it has the advantage of simplifying the units of the flux of this intensity, which is used a great deal.

We have now defined the electromagnetic field intensity vectors, and established units for their measurement, and for the measurement of charge. We must next consider the relationships between the field intensities and the charge distribution. This means we must consider a way to describe the motion of a charge distribution and to introduce the idea of a current.

---

* Called after A. G. A. A. Volta (1745–1827), Italian physicist. He studied electricity in animals and invented the first form of electric battery, the 'Voltaic Pile'.

† Called after W. E. Weber (1804–1890), German physicist. His main work was in the field of magnetic phenomena.

1.3. CURRENTS

If we have charged particles in motion, we say that an *electric current* is flowing. In order to describe more precisely the flow of charge in a system, we define a *current density* vector field $\mathbf{j}(\mathbf{r}, t)$ as follows. If we have a plane infinitesimal element of area $\mathbf{dS}$, which contains the point $\mathbf{r}$, then the amount of charge which crosses this area in an infinitesimal time dt is $\mathbf{j} \cdot \mathbf{dS} \, dt$. We must remember that the vector area $\mathbf{dS}$ has a magnitude equal to the area of the element dS and a direction normal to the surface. The rate of flow of charge across the element in the direction of the vector $\mathbf{dS}$ is $\mathbf{j} \cdot \mathbf{dS}$.

The unit of current, the ampère,* represents a flow of 1 coulomb of charge per second. The current density at any point in a system will in general be a function of the time.

It was established very early in the development of electro-magnetism that a current flowing in a wire gives rise to a magnetic field. We now know that any flow of charge gives rise to a magnetic field, and we must try and formulate a theory which will enable us to give a qualitative and quantitative description of the fields produced by systems of charges in motion.

1.4. THE FIELD EQUATIONS

We are trying to construct a theory, or mathematical model, which will enable us to describe both qualitatively and quantitatively those electromagnetic phenomena which are known from experiments, and also to predict further phenomena which may be verified experimentally. The way we shall construct the theory is to postulate relations between the physically measurable quantities and see whether the consequences of these postulates agree with experience. It is often thought to be easier to build up the relationships we shall postulate, the so-called *field equations*, by gradually generalising the results of simple experiments. However, even in this approach, the same or equivalent postulates must sooner or later be made, and they may as well be stated at the outset rather than trying to make them more plausible than their success in the correlation and prediction of phenomena indicates.

* Called after A. M. Ampère (1775–1836), one of the earliest workers in this subject. He was French and primarily an experimental physicist.

Suppose that we have a region of space in which we have an electromagnetic field as previously defined in Section 1.2. Let $\rho(\mathbf{r}, t)$ be the charge density in the region; i.e. if we take an element of volume $d\tau$ containing the point $\mathbf{r}$, then the amount of charge contained in this element at time t is $\rho(\mathbf{r}, t)\, d\tau$. Let $\mathbf{j}(\mathbf{r}, t)$ be the current density as defined in the previous section. Then for our initial postulates, we take the field equations of Maxwell,* i.e.

$$\nabla \times \mathbf{E} = -\frac{\partial \mathbf{B}}{\partial t}, \tag{1.4}$$

$$\nabla \times \mathbf{B} = \mu_0\left(\mathbf{j} + \epsilon_0\, \frac{\partial \mathbf{E}}{\partial t}\right), \tag{1.5}$$

$$\nabla . \mathbf{E} = \rho/\epsilon_0, \tag{1.6}$$

$$\nabla . \mathbf{B} = 0. \tag{1.7}$$

For completeness, we include the force on a charged particle, the Lorentz force, which defines the field intensities $\mathbf{E}$ and $\mathbf{B}$.

$$\mathbf{F} = q[\mathbf{E} + (\mathbf{v} \times \mathbf{B})]. \tag{1.8}$$

The constants ϵ_0 and μ_0 are defined by

$$\epsilon_0 = 8.854 \times 10^{-12} \text{ F/m}, \tag{1.9}$$
$$\mu_0 = 4\pi \times 10^{-7} \text{ henrys}\dagger/\text{m}.$$

It must be understood that these equations represent the assumptions on which our theory is based, and their suitability must be judged by the agreement with experiment of any results we derive, their self-consistency and their ability to predict new phenomena which can be experimentally verified. These are the main criteria by which we judge any theory or mathematical model.

PROBLEMS FOR CHAPTER 1

1. A particle of mass m and charge e describes a circle of radius r under the electric effect of a second charge e' at the centre of the circle. Determine the speed v of the charge.

* J. C. Maxwell (1831–1879), born in Scotland, and probably the greatest theoretical physicist of the nineteenth century.
† J. Henry (1797–1878), American physicist.

2. Show from the field equation

$$\mathbf{\nabla} \times \mathbf{E} = -\frac{\partial \mathbf{B}}{\partial t}$$

that if at any time **B** is zero then at any subsequent time the equation $\mathbf{\nabla} . \mathbf{B} = 0$ is satisfied.

3. Show that a magnetic field cannot increase the kinetic energy of a charged particle.

Chapter 2

THE FIELD EQUATIONS

In this chapter we shall begin to formulate some of the more important consequences of the field equations which were postulated in the last chapter.

We shall introduce the idea of representing an electromagnetic field by potentials, and then obtain these potentials as functions of the charge and current densities of the system. These results will lead us in a natural manner to consider how electromagnetic effects are propagated from one region of space to another, and will enable us to determine the velocity of propagation; or perhaps we should say that a characteristic velocity emerges which suggests a velocity of propagation. We shall return to this point in Chapter 5.

The first result to be deduced from the field equations is of great importance. It is the fact that the total charge in a system is constant. We must of course remember that each elementary charge in a system has a sign as well as a magnitude, and that the total charge in a system is the algebraic sum of these individual charges with due regard paid to their signs.

2.1. The Conservation of Charge

Let us take the divergence of the field equation (1.5) from the last section, and assume that we can interchange the order of the linear differential operators involved. The relation

$$\mathbf{\nabla}.(\mathbf{\nabla} \times \mathbf{B}) = \mu_0 \mathbf{\nabla}.\left(\mathbf{j} + \epsilon_0 \frac{\partial \mathbf{E}}{\partial t}\right),$$

in which the left-hand side is identically zero, reduces to

$$\mathbf{\nabla}.\mathbf{j} + \epsilon_0 \frac{\partial}{\partial t}(\mathbf{\nabla}.\mathbf{E}) = 0. \tag{2.1}$$

8

If we now combine equation (2.1) with the field equation (1.6), we obtain

$$\frac{\partial \rho}{\partial t} + \nabla \cdot \mathbf{j} = 0. \tag{2.2}$$

This is called the *equation of continuity of charge*. It expresses the fact that we cannot have net charge created or destroyed in any region of space. To see this result more clearly, let us integrate both sides of equation (2.2) over any fixed volume of space, say V. We write the result as

$$\iiint\limits_V \frac{\partial \rho}{\partial t}\, d\tau = -\iiint\limits_V \nabla \cdot \mathbf{j}\, d\tau,$$

where $d\tau$ is an element of volume of V. Let us now apply the divergence theorem to the right-hand side of this equation. Remembering that we are considering a fixed region of space, we obtain

$$\frac{\partial}{\partial t} \iiint\limits_V \rho\, d\tau = -\iint\limits_S \mathbf{j} \cdot d\mathbf{S}, \tag{2.3}$$

where S is the surface which encloses the region V; as usual the normal is in the outward sense. The left-hand side of equation (2.3) represents the rate of increase of charge contained in the volume V. The right-hand side of the equation represents the flux or total flow of charge through the boundary surface S, in the inward direction because of the negative sign outside the integral. Thus the amount of charge inside the volume V increases at just the rate at which charge flows into the volume through the boundary surface S; this fact implies that there is no net charge created in the region V. It would be possible for charge to be created, but for this it would be necessary to have equal and opposite amounts of positive and negative charge created, giving a zero net charge increase over the volume.

We have shown that the total charge in a system is constant, remembering that we must be careful of the signs of the component charges. This conclusion, which is drawn from our postulates, is in agreement with observations made on the macroscopic scale; it also agrees with observations on the decay of certain fundamental particles, which yield others having different individual charges.

2.2. Relationship between the Electric and Magnetic Fields

From the field equations (1.4) to (1.7) we can see that there is a close connection between time-dependent electric and magnetic fields. In fact in a system where the field intensities are time-dependent, we cannot have an electric field without a magnetic field, and *vice versa*. In a system where the intensities are constant it is possible to have the fields existing independently of each other; we shall return to these special cases in Chapters 3 and 4.

Let us now consider a phenomenon which arises when we have time-dependent magnetic flux through an open surface. This is now known as 'electromagnetic induction', discovered by Faraday: it is the basic principle on which electric generators are designed. Hence the following discussion is of great technological importance as well as being of theoretical significance.

Let us consider in a little more detail the consequences of the field equation

$$\nabla \times \mathbf{E} = -\frac{\partial \mathbf{B}}{\partial t}. \tag{2.4}$$

This equation implies, as we shall see after a little manipulation, that a time-dependent magnetic flux produces an electric field. This is the basis on which most machines for producing electric fields work. It is necessary to consider the process in two stages.

First let us suppose that a circuit of wire C is at rest in a time-dependent magnetic field $\mathbf{B}$. Then if S is a surface bounded by C we can integrate both sides of equation (2.4) over this surface to obtain

$$\iint_S (\nabla \times \mathbf{E}) . \mathbf{dS} = -\frac{\partial}{\partial t} \iint_S \mathbf{B} . \mathbf{dS}. \tag{2.5}$$

Here again we have assumed that we can put the differential operator outside the integral sign. The surface integral on the right-hand side of equation (2.5) is the flux of the magnetic intensity $\mathbf{B}$ over the surface S, say Φ. If we apply Stokes' Theorem* to the left-hand side of equation (2.5), we can write the equation in the form

$$\oint_C \mathbf{E} . \mathbf{ds} = -\frac{d}{dt} \iint_S \mathbf{B} . \mathbf{dS} = -\frac{d\Phi}{dt}, \tag{2.6}$$

since the circuit C is stationary.

---

* Sir G. G. Stokes (1819–1903), British mathematical physicist; he mainly worked on hydrodynamical problems.

We interpret equation (2.6) as follows. If we have a stationary circuit with a variable flux of magnetic intensity linking the circuit, then there is a force produced which will tend to drive a current in the circuit. Whether or not an actual current flows in the circuit depends on the availability of charged particles to produce the current. Any current produced flows in a direction such that the flux of its magnetic field through the circuit tries to compensate the change producing the current: this result is known as *Lenz's Law*.

Now let us suppose that the change in flux through a circuit C is caused by moving the circuit through a static magnetic field which is nonuniform (Fig. 1). Let S be the surface shown bounded by the circuit C at a time t. Let S' be a surface bounded by the circuit C

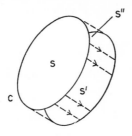

Fig. 1

at an infinitesimally later time $t + dt$. The loci of the points of the curve C will generate in this time dt a surface S''. Since the field equation $\nabla \cdot \mathbf{B} = 0$ is satisfied at all times, we know that the total flux of the magnetic intensity $\mathbf{B}$ over the surface made up of S, S' and S'' is zero. Hence if $d\Phi$ is the difference in the flux over S' and S, in the same sense, we have, since the field is static the flux over S is independent of time,

$$d\Phi = \text{flux of } \mathbf{B} \text{ over the surface } S''.$$

Let $\mathbf{ds}$ be an element of C at the point $\mathbf{r}$. In the infinitesimal time dt the element $\mathbf{ds}$ sweeps out an area $\mathbf{dS}$ given by

$$\mathbf{dS} = \mathbf{ds} \times \mathbf{v}(\mathbf{r}) \, dt,$$

where $\mathbf{v}(\mathbf{r})$ is the velocity of the element $\mathbf{ds}$ of C at $\mathbf{r}$, and the sense of $\mathbf{ds}$ is such that $\mathbf{dS}$ points outwards from the volume enclosed by S, S' and S''. The flux of magnetic induction over this element of area is given by

$$\mathbf{B} \cdot \mathbf{dS} = \mathbf{B} \cdot (\mathbf{ds} \times \mathbf{v}) \, dt.$$

For the total cylindrical surface S'' we obtain the flux of magnetic induction in the form

$$d\Phi = \int\int_{S''} \mathbf{B}.\mathbf{dS} = \oint_C \mathbf{B}.(\mathbf{ds} \times \mathbf{v})\, dt.$$

Hence the rate at which the flux of magnetic induction through S changes as S moves is given by

$$\frac{d\Phi}{dt} = \oint_C \mathbf{B}.(\mathbf{ds} \times \mathbf{v}) = -\oint_C (\mathbf{v} \times \mathbf{B}).\mathbf{ds}. \qquad (2.7)$$

Now any electron of charge e, which is negative, in the wire at the element $\mathbf{ds}$ experiences a force $\mathbf{F}$ given by

$$\mathbf{F} = e(\mathbf{v} \times \mathbf{B})$$

due to the motion of the wire. If we regard this as due to an electric field, say $\mathbf{E}'$, then equation (2.7) gives the relation

$$\oint_C \mathbf{E}'.\mathbf{ds} = -\frac{d\Phi}{dt}. \qquad (2.8)$$

This is of the same form as equation (2.6) for the stationary circuit. Thus again a current will tend to flow in the circuit. We call the line integrals on the left-hand sides of equations (2.6) and (2.8) *electromotive forces* (e.m.f.).

We have shown that whether the change in flux of magnetic induction through a circuit C is caused by a time-dependent magnetic field or by motion of the circuit in a static field, we obtain the same result, that the electromotive force produced round the circuit is equal to minus the rate of change of this flux. This electromotive force tries to produce an electric current in the circuit in a direction such that the magnetic flux produced by the current will compensate for the initial change in flux.

If we consider the more general case where the circuit C moves in a time-dependent magnetic field, then to obtain the rate of change of flux through C we must add the right-hand sides of equations (2.5) and (2.7). We obtain the law

$$\oint_C [\mathbf{E} + (\mathbf{v} \times \mathbf{B})].\mathbf{ds} = -\frac{d\Phi}{dt}. \qquad (2.9)$$

The left-hand side of this equation is the total electromotive force for the circuit, or the line integral of the force per unit charge on an electron round the circuit C.

The law relating the electromotive force induced in a circuit to the rate of change of magnetic flux through the circuit is known as *Faraday's Law of Induction*. It is one of the most widely applied results of the theory and (as we have said earlier) is of great technological importance. In practical devices for producing electricity, the e.m.f. is produced by causing a coil of wire to rotate in a static magnetic field. Then we have a so-called dynamo. Let us consider a simple example to illustrate the point.

▶ *Example*

A wire in the shape of a circle of radius r rotates about a diameter with constant angular velocity ω. There is a uniform magnetic field of intensity **B** normal to the axis of rotation. Find the e.m.f. produced in the coil at any moment.

▶ *Solution*

At any time t let θ be the angle between the normal to the plane of the wire loop and the direction of the magnetic intensity **B**. The flux of the magnetic intensity through the coil is equal to the magnetic intensity times the projection of the area of the coil in the direction of **B**, i.e.

$$\Phi = \pi r^2 B \cos\theta \text{ Wb}.$$

The rate of change of this flux due to the rotation of the coil is given by

$$\frac{d\Phi}{dt} = \frac{d}{dt}(B\pi r^2 \cos\theta) = -B\pi r^2 \sin\theta . \omega \text{ Wb/sec}.$$

Hence the e.m.f. produced in the wire loop at this time is given by

$$\text{e.m.f.} = -\frac{d\Phi}{dt} = B\pi r^2 \omega \sin\theta \text{ V}. \qquad ◀$$

2.3. POTENTIALS

The partial differential equations involving the field intensities, charge and current distributions that we have postulated are rather complicated, since they each contain more than one observable quantity. One problem that we are often confronted with in electromagnetic theory is that we are given the charge and current densities in a system and it is required that we obtain the electric and magnetic intensities as functions of them. This proves rather difficult with the field equations in the form we have at present. Before we can obtain a solution of the field equations we shall have to be given some boundary and initial conditions. However, even when these conditions are given, the solution of the field equations as we have them is difficult since the intensities are mixed up in each equation. We must find some method of simplifying the equations to the stage where each equation we try to solve has only one unknown quantity in it. One way of achieving this desired separation of the intensities is to introduce the concept of potential.

Consider the field equation

$$\mathbf{\nabla} \times \mathbf{E} = -\frac{\partial \mathbf{B}}{\partial t}. \tag{2.10}$$

Let us define a vector field $\mathbf{A}$ such that

$$\mathbf{B} = \mathbf{\nabla} \times \mathbf{A}. \tag{2.11}$$

This is always possible, and the number of choices for $\mathbf{A}$ is infinite. If we substitute from equation (2.11) into equation (2.10) for $\mathbf{B}$, and assume that we can interchange the order of the operators, we obtain

$$\mathbf{\nabla} \times \left(\mathbf{E} + \frac{\partial \mathbf{A}}{\partial t} \right) = 0. \tag{2.12}$$

Now any vector which has a vanishing curl can be written as the gradient of a scalar field, and so we write

$$\mathbf{E} + \frac{\partial \mathbf{A}}{\partial t} = -\mathbf{\nabla}\phi. \tag{2.13}$$

This scalar function ϕ that we have introduced is called the *electric potential*, for reasons we shall appreciate better at a later stage, and the vector $\mathbf{A}$ is known as the *magnetic vector potential*.

The potential functions we have introduced are not unique. For instance, to $\mathbf{A}$ we could add the gradient of any scalar field, since this has a zero curl, and equation (2.11) would still be satisfied. We shall have more to say on this question of uniqueness later.

In terms of the potentials the field intensities have the following form:

$$\left.\begin{aligned} \mathbf{B} &= \mathbf{\nabla} \times \mathbf{A} \\ \mathbf{E} &= -\mathbf{\nabla}\phi - \frac{\partial \mathbf{A}}{\partial t} \end{aligned}\right\}. \tag{2.14}$$

If we can solve the problem of obtaining the potentials ϕ and $\mathbf{A}$ as functions of the charge and current densities ρ and $\mathbf{j}$, then we can derive the intensities from them by using equations (2.14). We shall see that this is often an easier procedure than the direct solution of the field equations for the intensities $\mathbf{E}$ and $\mathbf{B}$ in terms of ρ and $\mathbf{j}$.

Using the representation (2.14) in the field equation

$$\mathbf{\nabla} \times \mathbf{B} = \mu_0\left(\mathbf{j} + \epsilon_0 \frac{\partial \mathbf{E}}{\partial t}\right)$$

gives

$$\mathbf{\nabla} \times (\mathbf{\nabla} \times \mathbf{A}) = \mu_0\left[\mathbf{j} + \epsilon_0 \frac{\partial}{\partial t}\left(-\mathbf{\nabla}\phi - \frac{\partial \mathbf{A}}{\partial t}\right)\right]. \tag{2.15}$$

We again assume that we can interchange the operators, and thus use the fact that

$$\mathbf{\nabla} \times (\mathbf{\nabla} \times \mathbf{A}) = \mathbf{\nabla}(\mathbf{\nabla} \cdot \mathbf{A}) - \nabla^2 \mathbf{A}$$

to obtain

$$\nabla^2 \mathbf{A} - \mu_0 \epsilon_0 \frac{\partial^2 \mathbf{A}}{\partial t^2} - \mathbf{\nabla}\left(\mathbf{\nabla} \cdot \mathbf{A} + \mu_0 \epsilon_0 \frac{\partial \phi}{\partial t}\right) = -\mu_0 \mathbf{j}. \tag{2.16}$$

This equation still contains both potentials and as yet we have not achieved the separation we desire. If however we use the representation (2.14) in the equation

$$\mathbf{\nabla} \cdot \mathbf{E} = \rho/\epsilon_0,$$

we obtain

$$\nabla^2 \phi + \frac{\partial}{\partial t}(\mathbf{\nabla} \cdot \mathbf{A}) = -\frac{\rho}{\epsilon_0}.$$

We shall write this, for reasons we shall soon discuss, in the more complicated form

$$\nabla^2 \phi - \mu_0 \epsilon_0 \frac{\partial^2 \phi}{\partial t^2} + \frac{\partial}{\partial t}\left(\mathbf{\nabla} \cdot \mathbf{A} + \mu_0 \epsilon_0 \frac{\partial \phi}{\partial t}\right) = -\frac{\rho}{\epsilon_0}. \tag{2.17}$$

This equation again contains both potentials, and we have so far not simplified the problem at all. However, we did mention the fact that the potentials were not uniquely specified yet. A vector field is only completely specified when we give at each point of space its curl and divergence, and when we also give its behaviour on the surface bounding its region of definition (if the region is unbounded then we have to give the behaviour at infinity). This result is known as *Helmholtz's Theorem.*[*] We have specified the curl of the vector **A** by equation (2.11). We now subject the divergence of **A** to the restriction

$$\mathbf{\nabla} \cdot \mathbf{A} + \mu_0 \epsilon_0 \frac{\partial \phi}{\partial t} = 0, \tag{2.18}$$

which is known as the *Lorentz condition*. However, **A** is still not uniquely defined, a point we shall again return to later. Now we can achieve some simplification of the equations containing the potentials. Using the Lorentz condition (2.18) in equations (2.16) and (2.17), we find that separation of the potentials occurs and that each potential satisfies an inhomogeneous wave equation. We obtain

$$\mathbf{\nabla}^2 \mathbf{A} - \frac{1}{c^2} \frac{\partial^2 \mathbf{A}}{\partial t^2} = -\mu_0 \mathbf{j}, \tag{2.19}$$

$$\mathbf{\nabla}^2 \phi - \frac{1}{c^2} \frac{\partial^2 \phi}{\partial t^2} = -\frac{\rho}{\epsilon_0}. \tag{2.20}$$

Here we have written

$$c = (\mu_0 \epsilon_0)^{-1/2}.$$

We see that the concept of potential has allowed us to reduce the number of unknowns on the left-hand side of each equation to one; although this is at the expense of increasing the order of the equations from one to two, it usually simplifies the problem. We shall consider the general solutions of equations (2.19) and (2.20) in the next section.

The constant c in equations (2.19) and (2.20) has the dimensions of a velocity. It has a magnitude of approximately 3×10^8 m/sec. We shall discuss the physical significance of this velocity later.

Let us for a moment return to the question of the non-uniqueness of the potentials ϕ and **A** as we have defined them. If we consider ϕ_0

---

[*] H. L. F. von Helmholtz (1821–1894), the great German physicist. He is noted for his work in acoustics and his formulation of the principle of conservation of energy.

and $\mathbf{A}_0$ to be a set of potentials which lead to the given field intensities when inserted in equation (2.14), then the same result is obtained with the set ϕ and $\mathbf{A}$ defined by

$$\phi = \phi_0 - \frac{\partial \psi}{\partial t},$$

$$\mathbf{A} = \mathbf{A}_0 - \nabla \psi,$$

(2.21)

where ψ is any solution of the homogeneous wave equation

$$\nabla^2 \psi - \frac{1}{c^2} \frac{\partial^2 \psi}{\partial t^2} = 0.$$

(2.22)

The new potentials ϕ and $\mathbf{A}$ also satisfy the Lorentz condition (2.18) if the potentials ϕ_0 and $\mathbf{A}_0$ do. We refer to the transformations of the form (2.21) as *gauge transformations*. We must make sure that any physically observable consequence of our theory which is expressed in terms of the potentials is invariant under these changes of potential, or as we say must be *gauge invariant*, since the intensities which are the directly measurable quantities of the theory are invariant under such transformations. If we had an observable quantity expressed in terms of the potentials and it was not gauge invariant, it would be a quantity which was not uniquely defined for a given field configuration. This would be meaningless in a classical theory. Hence any combination of potentials which is not gauge invariant cannot correspond to a physically observable quantity.

2.4. LINEAR ASPECTS OF THE THEORY

The basic differential equations (2.19) and (2.20) satisfied by the potentials are in a certain sense linear equations. By this we mean that if a charge and current distribution with densities ρ_1 and $\mathbf{j}_1$ give rise to a field with potentials ϕ_1 and $\mathbf{A}_1$, and a distribution ρ_2 and $\mathbf{j}_2$ gives a field with potentials ϕ_2 and $\mathbf{A}_2$, then the distribution with densities $\rho_1 + \rho_2$ and $\mathbf{j}_1 + \mathbf{j}_2$ gives a field with potentials $\phi_1 + \phi_2$ and $\mathbf{A}_1 + \mathbf{A}_2$. Thus this type of linearity of the basic equations for the potentials means that we can add solutions of the equations corresponding to given densities of charge and current. Since the field intensities are derived from the potentials by the application of *linear* operators, we see that the intensities also possess the same property.

Thus we see that if we have two known distributions of charge and current and if we add the intensities at any point, the resultant corresponds to the intensity for a distribution which is the sum of the two original distributions. We call this result the *principle of superposition*.

A linear theory, or a theory in which a principle of superposition is valid, has considerable advantages over a nonlinear theory where no such principle exists. In the former a problem involving a complicated system can often be solved in a piecemeal fashion by breaking the system into components and then adding the results for each component in some suitable manner. In a nonlinear theory one cannot reduce the study of a compound system to a study of its components, but must always consider the system as a whole. There is also a reasonable theory for solving linear partial differential equations, whereas in the nonlinear case each equation must be treated separately on its own merits and the problem of obtaining a solution is generally much more difficult. So indeed we are fortunate that the theory of the electromagnetic field is at least linear in the sense we have described above.

We must be very careful not to think that the theory of the electromagnetic field is linear in all aspects. We shall see in the final chapter of this work that when one considers the motion of a single charged particle in detail then it may be (we say 'may be' because the problem has not yet been completely solved) that these basic equations are nonlinear. In most macroscopic applications of the theory we can consider the charge and current densities to be prescribed and then we can use the linear property we have discussed above. If one considers the electromagnetic field in media, then this also may introduce nonlinear effects into the theory and one loses the simplicity again.

Now let us turn to the problem of obtaining the solutions of the inhomogeneous wave equations which are satisfied by the potentials. We shall solve the equations in the most general form and return to the special cases of static charge and current densities later.

2.5. The Solution of the Inhomogeneous Wave Equations

We have shown that the field intensities $\mathbf{E}$ and $\mathbf{B}$ can be derived from potentials ϕ and $\mathbf{A}$ which satisfy inhomogeneous wave equations.

The method of solution of these equations which we shall consider is due to Kirchhoff.*

The equations satisfied by the electric potential ϕ and each component of the magnetic vector potential $\mathbf{A}$ are of the form

$$\mathbf{V}^2 V - \frac{1}{c^2}\frac{\partial^2 V}{\partial t^2} = -g(\mathbf{r}, t). \tag{2.23}$$

We wish to obtain the function $V(\mathbf{r}, t)$ as a function of $g(\mathbf{r}, t)$, $\mathbf{r}$ and t.

We begin by using an integral transform to change equation (2.23) into a more manageable form. Using results from the theory of Fourier† integrals, we can write the functions $V(\mathbf{r}, t)$ and $g(\mathbf{r}, t)$ in the form

$$\left.\begin{aligned}
g(\mathbf{r}, t) &= \int\limits_{-\infty}^{+\infty} \bar{g}(\mathbf{r}, p)\, e^{ipt}\, dp \\
V(\mathbf{r}, t) &= \int\limits_{-\infty}^{+\infty} \bar{V}(\mathbf{r}, p)\, e^{ipt}\, dp
\end{aligned}\right\}, \tag{2.24}$$

with the functions $\bar{g}(r, p)$ and $\bar{V}(r, p)$ given by the relations

$$\left.\begin{aligned}
\bar{g}(\mathbf{r}, p) &= \frac{1}{2\pi} \int\limits_{-\infty}^{+\infty} g(\mathbf{r}, t)\, e^{-ipt}\, dt \\
\bar{V}(\mathbf{r}, p) &= \frac{1}{2\pi} \int\limits_{-\infty}^{+\infty} V(\mathbf{r}, t)\, e^{-ipt}\, dt
\end{aligned}\right\}. \tag{2.5}$$

We call the function $\bar{g}(\mathbf{r}, p)$ the *Fourier transform* of the function $g(\mathbf{r}, t)$. We shall assume that in the following we are always dealing with functions which have a Fourier transform, i.e. that the integrals in (2.25) actually exist.

If we use the representation (2.24) in equation (2.23), and assume that we can differentiate under the integral sign, we obtain the equation

$$\int\limits_{-\infty}^{+\infty} (\mathbf{V}^2 \bar{V} + k^2 \bar{V})\, e^{ipt}\, dp = -\int\limits_{-\infty}^{+\infty} \bar{g}\, e^{ipt}\, dp,$$

---

* G. R. Kirchhoff (1824–1887), German physicist noted for his pioneer work in spectroscopy.

† Baron J. B. J. Fourier (1768–1830). He introduced the subject now known as Fourier analysis in his researches on heat flow.

where we have introduced the new constant k given by

$$k = p/c.$$

We rewrite this transformed equation in the form

$$\int_{-\infty}^{+\infty} (\nabla^2 \overline{V} + k^2 \overline{V} + \overline{g}) \, e^{ipt} \, dp = 0.$$

However, we can see from equations (2.25) that the Fourier transform of a function is zero when the function itself is zero. We thus write this last equation in the form

$$\nabla^2 \overline{V} + k^2 \overline{V} = -\overline{g}, \tag{2.26}$$

which is an easier equation to solve than (2.23). This simplification was the reason for the use of Fourier transforms.

We shall now solve equation (2.26) by the Green's function* method which we consider to be the most suitable. Let $\mathbf{r}$ be the point of space at which we wish to obtain the solution of equation (2.26) at time t. Consider a function $G(\mathbf{r}, \mathbf{r}', p)$ which satisfies the equation

$$\nabla'^2 G + k^2 G = 0 \quad (\mathbf{r}' \neq \mathbf{r}) \tag{2.27}$$

where the prime denotes differentiation with respect to $\mathbf{r}'$. The particular function we shall consider which has this property is

$$G(\mathbf{r}, \mathbf{r}', p) = \frac{e^{-ik|\mathbf{r}-\mathbf{r}'|}}{|\mathbf{r}-\mathbf{r}'|} \tag{2.28}$$

with k defined as above. Now let us consider the dependence of this function on $\mathbf{r}'$. The vector $\mathbf{r}'$ is used to denote the position of points in space other than the *field point* $\mathbf{r}$. Suppose we surround the point $\mathbf{r}$ by a small sphere S_0, of radius ϵ, with its centre at $\mathbf{r}$. Let S be some closed surface which completely contains S_0, as shown in Fig. 2. Let V' be the region between the surfaces S_0 and S. In vector analysis we have a result known as *Green's Second Theorem* which states that if we have two scalar functions ϕ and ψ defined in a region D with their first derivatives continuous and single valued in D, and with $\nabla^2\phi$, $\nabla^2\psi$ integrable over D, then if S is the boundary of D,

$$\iiint_D (\phi\nabla^2\psi - \psi\nabla^2\phi) \, d\tau = \iint_S (\phi\nabla\psi - \psi\nabla\phi) \cdot d\mathbf{S}, \tag{2.29}$$

---

* G. Green (1793–1841), British mathematician. He did great work in potential theory.

where in the surface integral the positive normal is in the outward sense from the region D.

Consider the special application of equation (2.29) in which for ϕ we take our function G defined by equation (2.28), for ψ we take $\overline{V}(\mathbf{r}, p)$ defined by (2.25), and we take all differentiations with respect to $\mathbf{r}'$. For the region D we take the region V' between S_0 and S. In this case we obtain the result

$$\iiint\limits_{V'} (G\nabla'^2\overline{V} - \overline{V}\nabla'^2 G) \, d\tau = \iint\limits_{S+S_0} (G\nabla'\overline{V} - \overline{V}\nabla'G).\mathbf{dS}. \qquad (2.30)$$

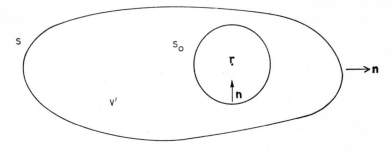

FIG. 2

For the surface integrals on the right-hand side the positive normal is inwards from S_0 and outwards from S, as shown in Fig. 2. It is easier to study equation (2.30) if we separate some of the terms and write it in the form

$$\iiint\limits_{V'} (G\nabla'^2\overline{V} - \overline{V}\nabla'^2 G) \, d\tau$$

$$= \iint\limits_{S} (G\nabla'\overline{V} - \overline{V}\nabla'G).\mathbf{dS} + \iint\limits_{S_0} G\nabla'\overline{V}.\mathbf{dS} - \iint\limits_{S_0} \overline{V}\nabla'G.\mathbf{dS} \qquad (2.31)$$

Consider the second term on the right-hand side of this equation. We assume that $\nabla'\overline{V}$ is finite. If this was not so, then from equation (2.24) we would have $\nabla'V$ infinite and this would correspond to an infinite field intensity, which is not physically realistic. We are leaving aside for the moment the problem of determining the field at a point $\mathbf{r}$ which is occupied by a point charge, where the assumption that we make breaks down. We shall see later that while the potential of an isolated point charge can be assumed to have the

3

form r^{-1} for values of r which are not too small, if we try to extend this to all values of r we encounter severe difficulties in the physical interpretation.

As the radius of the sphere S_0, which we have written as ϵ, tends to zero, the surface area of S_0 is of the order of ϵ^2. The function G is of the order of ϵ^{-1} as ϵ tends to zero; since we have assumed the term $\nabla' \overline{V}$ bounded, we have that this complete surface integral tends to zero as ϵ tends to zero. Thus if we are interested in the limiting form of equation (2.31) as ϵ tends to zero, then we can drop the second term on the right-hand side, bearing in mind the restrictions we have outlined above.

Now consider the last term on the right-hand side of equation (2.31). As ϵ tends to zero this term tends to the following expression

$$\underset{\epsilon \to 0}{\text{Lt}} \iint_{S_0} \overline{V}(\mathbf{r}') \left\{ \frac{-e^{-ik|\mathbf{r}-\mathbf{r}'|}}{|\mathbf{r}-\mathbf{r}'|^2} - \frac{ik\, e^{-ik|\mathbf{r}-\mathbf{r}'|}}{|\mathbf{r}-\mathbf{r}'|} \right\} \frac{(\mathbf{r}-\mathbf{r}')}{|\mathbf{r}-\mathbf{r}'|} . \mathbf{dS}.$$

If we remember that the element of surface area behaves as the square of the radius of the sphere, this reduces to

$$- \underset{\epsilon \to 0}{\text{Lt}} \iint_{S_0} \overline{V}(\mathbf{r}') \frac{(\mathbf{r}-\mathbf{r}')}{|\mathbf{r}-\mathbf{r}'|^3} . \frac{(\mathbf{r}-\mathbf{r}')}{|\mathbf{r}-\mathbf{r}'|} \, dS = -4\pi \overline{V}(\mathbf{r}). \qquad (2.32)$$

Hence in the limit as ϵ tends to zero, equation (2.31) becomes

$$-4\pi \overline{V}(\mathbf{r}) = \iiint_{V'} (G\nabla'^2 \overline{V} - \overline{V}\nabla'^2 G) \, d\tau - \iint_{S} (G\nabla' \overline{V} - \overline{V}\nabla' G) . \mathbf{dS}. \qquad (2.33)$$

We see that on the left-hand side we have the quantity we seek, the value of the function $\overline{V}$ at the point $\mathbf{r}$. We now try to simplify the right-hand side of the equation. The final function that we seek is actually not $\overline{V}(\mathbf{r}, p)$ but $V(\mathbf{r}, t)$, but this is easily obtained from $\overline{V}$ by inverting the Fourier transform by means of equations (2.24).

From equations (2.26) and (2.27) we have

$$G\nabla'^2 \overline{V} - \overline{V}\nabla'^2 G = -\bar{g}G,$$

which on substitution into equation (2.33) gives

$$4\pi \overline{V}(\mathbf{r}) = \iiint_{V'} \bar{g}G \, d\tau + \iint_{S} \left(G\, \frac{\partial \overline{V}}{\partial n} - \overline{V}\, \frac{\partial G}{\partial n} \right) dS. \qquad (2.34)$$

Here $\partial/\partial n$ denotes the normal derivative to the surface S in the outward sense. We can expand the term $\partial G/\partial n$ which appears in this equation to obtain

$$\frac{\partial G}{\partial n} = \left[\frac{-ik}{|\mathbf{r}-\mathbf{r}'|} \frac{\partial |\mathbf{r}-\mathbf{r}'|}{\partial n} + \frac{\partial}{\partial n}\left(\frac{1}{|\mathbf{r}-\mathbf{r}'|}\right) \right] e^{-ik|\mathbf{r}-\mathbf{r}'|}. \qquad (2.35)$$

If we substitute for $\partial G/\partial n$ from equation (2.35) into equation (2.34), we obtain

$$4\pi \overline{V}(\mathbf{r}, p) = \iiint_{V'} \bar{g}(\mathbf{r}', p) G(\mathbf{r}, \mathbf{r}', p)\, d\tau$$
$$+ \iint_{S} \left[\frac{\partial \overline{V}}{\partial n} \frac{1}{|\mathbf{r}-\mathbf{r}'|} + \frac{ik\overline{V}}{|\mathbf{r}-\mathbf{r}'|} \frac{\partial |\mathbf{r}-\mathbf{r}'|}{\partial n} \right.$$
$$\left. - \overline{V} \frac{\partial}{\partial n}\left(\frac{1}{|\mathbf{r}-\mathbf{r}'|}\right) \right] e^{-ik|\mathbf{r}-\mathbf{r}'|}\, dS.$$

This gives us the Fourier transform $\overline{V}(\mathbf{r}, p)$ of the solution we seek, $V(\mathbf{r}, t)$. Let us now invert this transform to obtain the actual solution. Using the inversion formula (2.24) we obtain, inverting the order of integration,

$$4\pi V(\mathbf{r}, t) = \iiint_{V'} \left[\int_{-\infty}^{+\infty} \frac{\bar{g}(\mathbf{r}, p)\, e^{ip(t-|\mathbf{r}-\mathbf{r}'|/c)}}{|\mathbf{r}-\mathbf{r}'|}\, dp \right] d\tau$$
$$+ \iint_{S} \left\{ \int_{-\infty}^{+\infty} \left[\frac{1}{|\mathbf{r}-\mathbf{r}'|} \frac{\partial \overline{V}}{\partial n} + \frac{ik\overline{V}}{|\mathbf{r}-\mathbf{r}'|} \frac{\partial |\mathbf{r}-\mathbf{r}'|}{\partial n} \right. \right.$$
$$\left. \left. - \overline{V} \frac{\partial}{\partial n}\left(\frac{1}{|\mathbf{r}-\mathbf{r}'|}\right) \right] e^{ip(t-|\mathbf{r}-\mathbf{r}'|/c)}\, dp \right\} dS. \qquad (2.36)$$

However, we can write this in a more compact form if we notice that the effect of the exponential terms is to change the argument of the inverted transform. We put

$$4\pi V(\mathbf{r}, t) = \iiint_{V'} \frac{[g]}{|\mathbf{r}-\mathbf{r}'|}\, d\tau$$
$$+ \iint_{S} \left\{ \frac{1}{|\mathbf{r}-\mathbf{r}'|}\left[\frac{\partial V}{\partial n}\right] + \frac{1}{c}\left[\frac{\partial V}{\partial t}\right] \frac{1}{|\mathbf{r}-\mathbf{r}'|} \frac{\partial |\mathbf{r}-\mathbf{r}'|}{\partial n} \right.$$
$$\left. - [V] \frac{\partial}{\partial n}\left(\frac{1}{|\mathbf{r}-\mathbf{r}'|}\right) \right\} dS. \qquad (2.37)$$

Here square brackets round a quantity mean that it is to be evaluated at the retarded time T defined by

$$T = t - |\mathbf{r} - \mathbf{r}'|/c. \tag{2.38}$$

Of course, as we can see, equation (2.37) does not give us the value of $V(\mathbf{r}, t)$ explicitly. It merely gives us this value as a function of the values of V on the surface S, and these are not arbitrary. We can however make some progress as follows.

Suppose we imagine that all sources of the electromagnetic field are located in a finite volume D, and that these sources were established within a finite time of the instant the field is measured at $\mathbf{r}$. If we assume that electromagnetic disturbances can only propagate *from* sources with a finite velocity of propagation, a point we shall return to in more detail later, then we can fix the surface S referred to in our equation (2.37) so far from the region D that any disturbance will not have had time to reach it. This of course means that the integrand of the surface integral will be zero, and hence we can remove this term. If the assumptions we have made are justified, again a point which must be judged on the success of the results derived, we obtain for the function $V(\mathbf{r}, t)$ the representation

$$V(\mathbf{r}, t) = \frac{1}{4\pi} \int \int \int \frac{[g]}{|\mathbf{r} - \mathbf{r}'|} \, d\tau, \tag{2.39}$$

where the integration is over the total volume of space in which the source function g is nonzero at the retarded time T.

If we now relate the solution (2.39) of the equation (2.23) to the original problem of determining the potentials which satisfy equations (2.19) and (2.20), we see that

$$\phi(\mathbf{r}, t) = \frac{1}{4\pi\epsilon_0} \int \int \int \frac{[\rho]}{|\mathbf{r} - \mathbf{r}'|} \, d\tau, \tag{2.40}$$

$$\mathbf{A}(\mathbf{r}, t) = \frac{\mu_0}{4\pi} \int \int \int \frac{[\mathbf{j}]}{|\mathbf{r} - \mathbf{r}'|} \, d\tau, \tag{2.41}$$

with the volumes of integration as defined in the previous paragraph. The expressions given in (2.40) and (2.41) are known as the *retarded potentials* for the inhomogeneous wave equation, for obvious reasons. Let us consider their meaning in a little more detail.

Suppose we wish to consider the potential of the field at a point $\mathbf{r}$ at a time t. Then the contribution to this potential from the volume element $d\tau$ at the point $\mathbf{r}'$ is not governed by the source density at $\mathbf{r}'$ at the time t but by the source density at the retarded time T defined by equation (2.38). Thus we can imagine that the contribution to the potential at $\mathbf{r}$ at time t, and hence the contribution to the intensities, are propagated from $\mathbf{r}'$ to $\mathbf{r}$ with a speed of propagation equal to c. They leave the source point $\mathbf{r}'$ at the retarded time T and arrive at the field point $\mathbf{r}$ at the time t at which the measurement is made. This idea of a finite speed of propagation is in accord with the assumptions made earlier to arrive at equation (2.39) from equation (2.37).

If we had chosen the Green's function

$$G(\mathbf{r}, \mathbf{r}', p) = \frac{e^{ik|\mathbf{r}-\mathbf{r}'|}}{|\mathbf{r}-\mathbf{r}'|}$$

instead of that given by equation (2.28), we should have arrived at a solution of the inhomogeneous wave equation which was similar to that given by (2.31), but where the functions had to be evaluated at a time T' which is referred to as the advanced time, and is given by

$$T' = t + |\mathbf{r}-\mathbf{r}'|/c. \tag{2.42}$$

The solutions for ϕ and $\mathbf{A}$ are then referred to as the *advanced potentials*.

While the advanced potentials are perfectly valid representations from the mathematical point of view, it is generally regarded as obvious that from the physical point of view such solutions must be rejected. The use of advanced potentials would imply that the contribution to the potential at the point $\mathbf{r}$ at time t was due to the source density at $\mathbf{r}'$ at time T' given by equation (2.42). This seems to contradict the ideas of causality which most people believe to be valid, namely that events can be separated into cause and effect and that cause precedes effect. Of course it may be that the division of events into cause and effect is the reason for our conceptual difficulties over the use of advanced potentials.

The advanced potentials were ignored for a long time, for the reasons we have discussed; but they are now receiving a great deal of attention, since they seem to be a means of avoiding some of the difficulties which beset classical electromagnetic theory. They were first invoked to try and solve the problem of obtaining an equation of motion for an electron moving in an electromagnetic field which would take into consideration the radiation reaction, which is the force which acts on an electron due to its own electromagnetic field. Attempts to solve this problem using only retarded potentials have proved unsuccessful. There are other problems concerned with the structure and stability of charged particles where the advanced potentials also give some hope of success. However, unless otherwise stated we shall confine ourselves in this work to the study of the theory based on the retarded potentials.

In this chapter we have solved the equations for the potentials in the most general case when the behaviour of the field is time-dependent. In the next two chapters we turn our attention to the very special, but important, cases when we have time-independent charge and current densities.

PROBLEMS FOR CHAPTER 2

(1) Show that if in any region of space the current density $\mathbf{j}$ is related to the electric intensity by $\mathbf{j} = \sigma \mathbf{E}$, then any charge present decreases in density exponentially with time (σ a positive constant).

(2) Show that if at any time the electric intensity and charge density are zero, then the equation $\nabla \cdot \mathbf{E} = \rho/\epsilon_0$ follows from the equation of continuity of charge.

(3) Show that in a region of space free from charge and currents we could represent the field intensities by a scalar potential ψ and a vector potential $\mathbf{T}$ with

$$\mathbf{B} = -\nabla\psi + \frac{1}{c^2}\frac{\partial \mathbf{T}}{\partial t}, \qquad \mathbf{E} = \nabla \times \mathbf{T}.$$

(4) Derive the advanced potential solutions of the inhomogeneous wave equation.

The author is indebted to G. D. Wassermann for the method of solution of equation (2.23) which results in equation (2.37).

CHAPTER 3

THE ELECTROSTATIC FIELD

In this chapter we shall be concerned with systems in which the charge distribution is static, residing on fixed bodies. The general expressions that we have obtained for the field intensities and potentials become very much simplified by the restriction to static systems.

3.1. THE STATIC ELECTRIC FIELD

In a system of static charges, we can put the current density equal to zero. We can also equate to zero all those terms in the field equations which involve a differentiation with respect to time, since we expect that a static system of charges will give rise to a static electromagnetic field. We find the field equations break down into two sets, one for the magnetic intensity and one for the electric intensity. They become

$$\mathbf{\nabla}.\mathbf{E} = \rho/\epsilon_0, \qquad (3.1)$$

$$\mathbf{\nabla} \times \mathbf{E} = 0, \qquad (3.2)$$

and

$$\mathbf{\nabla}.\mathbf{B} = 0, \qquad (3.3)$$

$$\mathbf{\nabla} \times \mathbf{B} = 0. \qquad (3.4)$$

Since the magnetic field $\mathbf{B}$ must certainly vanish at very large distances from the system of charges, equations (3.3) and (3.4) when considered with Helmholtz's equation imply that the magnetic intensity $\mathbf{B}$ is everywhere zero. Thus we are left with an electric field $\mathbf{E}$.

The potential representation of this field takes the form

$$\mathbf{E} = -\mathbf{\nabla}\phi. \qquad (3.5)$$

27

Here the electric scalar potential ϕ satisfies the Poisson* equation

$$\nabla^2\phi = -\rho/\epsilon_0, \tag{3.6}$$

which has the solution

$$\phi(\mathbf{r}) = \frac{1}{4\pi\epsilon_0} \int\int\int \frac{\rho(\mathbf{r}')}{|\mathbf{r}-\mathbf{r}'|} \, d\tau, \tag{3.7}$$

where the volume integral extends over the total charge distribution. These equations (3.5), (3.6) and (3.7) are the special cases of equations (2.14), (2.20) and (2.40) respectively corresponding to the static conditions where $[\rho(\mathbf{r}')] = \rho(\mathbf{r}')$. Thus the potential at any point is the sum of constant contributions from each fixed volume element.

If we are given the charge density of the system and we wish to determine the electric intensity, we can either solve equation (3.6), subject to certain boundary conditions, or perform the integration in equation (3.7). The way we choose to proceed depends on the problem in hand and on the information given. For instance, if we are given $\rho(\mathbf{r})$ in a region of space V and the value of the potential ϕ on the boundary S of V, then to find the value of ϕ at all points of V we would use equation (3.6). If however we are given $\rho(\mathbf{r})$ at all points in space, then we would use equation (3.7). Having obtained the potential, we obtain the electric intensity of the electric field by using equation (3.5).

In many cases, particularly if the charge distribution has any degree of symmetry, it is easier to obtain the electric intensity from an integral form of the field equations, which we shall now consider. Consider a region V bounded by a closed surface S. Integrate both sides of equation (3.1) over this region, giving

$$\int\int\int_V \nabla.\mathbf{E} \, d\tau = \frac{1}{\epsilon_0} \int\int\int_V \rho \, d\tau. \tag{3.8}$$

The integral on the right-hand side of this equation is merely the total charge contained in the region V, say Q. If we use the divergence theorem, we can write this equation in the form

$$\int\int_S \mathbf{E}.\mathbf{dS} = Q/\epsilon_0. \tag{3.9}$$

If we can find a surface over which $\mathbf{E}$ is either tangential, or normal

---

* S. D. Poisson (1781–1840), French mathematician, noted for his work on definite integrals and probability.

and of constant magnitude (this may be deduced from symmetry), we can write equation (3.9) as

$$EA = Q/\epsilon_0$$

or

$$E = Q/\epsilon_0 A, \tag{3.10}$$

where A is the area of the portion of the surface over which $\mathbf{E}$ is normal and of constant magnitude E. In this approach to obtaining the electric potential a fairly high degree of symmetry of the charge distribution is essential. In a moment we shall do an example of this type to illustrate the use of this procedure.

We shall, in this chapter, consider briefly examples of obtaining the field intensity for systems of static charges by each of the three methods we have outlined above:

(i) using equation (3.7);
(ii) using equation (3.6) with boundary conditions; and
(iii) using equation (3.10).

The last method being the simplest, let us consider it first.

▶ *Example*

Determine the electric intensity of the field due to a point charge.

▶ *Solution*

Choose a co-ordinate system so that the charge is at the origin. Consider a spherical surface with radius r and centre at the origin. By the symmetry of the system, the electric intensity must be normal and constant over this surface. From equation (3.10) we obtain

$$4\pi r^2 E(r) = Q/\epsilon_0.$$

Hence

$$E(r) = Q/4\pi\epsilon_0 r^2.$$

Finally we can write

$$\mathbf{E(r)} = Q\mathbf{r}/4\pi\epsilon_0 r^3,$$

and we have determined the required electric intensity. ◀

From the solution of this problem we can illustrate *Coulomb's Law* for the electric force between two charged particles at rest. In many treatments of the subject, this forms the starting-point of discussions on the electrostatic field. Coulomb's Law states that the electric force between two charged particles at rest, charges q and q' let us say, distance d apart *in vacuo* is $qq'/4\pi\epsilon_0 d^2$. We have of course already used this to define the unit of charge in Section 1.1.

Now let us consider a problem which is slightly more complex, although it still shows a high degree of symmetry.

▶ *Example*

Determine the electric intensity of a spherically symmetric distribution of charge.

▶ *Solution*

Choose the origin of our co-ordinate system to be at the centre of symmetry. Let $\rho(r)$ be the charge density at a distance r from the origin. Take a sphere V of radius r, centre the origin, and apply equation (3.10) to its surface S. As in the previous example, we see from the symmetry that the intensity is normal to S and constant on S. Thus we obtain

$$4\pi r^2 E(r) = \frac{1}{\epsilon_0} \iiint\limits_V \rho(r) \, d\tau.$$

If we use spherical polar co-ordinates this reduces to

$$E(r) = \frac{1}{\epsilon_0 r^2} \int_0^r \rho(p) p^2 \, dp,$$

or, finally

$$\mathbf{E(r)} = \frac{\mathbf{r}}{\epsilon_0 r^3} \int_0^r \rho(p) p^2 \, dp. \qquad\qquad ◀$$

The next example we shall consider is the electric dipole. This is a very important concept used in the study of the effects of electromagnetic fields in matter. It forms the basis for simple models to explain 'polarisation' in dielectrics and 'magnetisation' in metals. We shall not be concerned with these phenomena in this work, but merely mention them in order to explain the importance of the results of the following problem.

▶ *Example*

As a simple model of an electric dipole, let us consider the field of two equal and opposite charges *in vacuo*.

▶ *Solution*

Let us consider a positive charge $+e$ at the point A and a negative charge $-e$ at the point B, with $AB = 2d$. Suppose we wish to find the electric intensity at a point P (Fig. 3). Consider

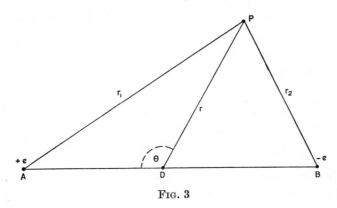

FIG. 3

the plane through A, B and P. Since the system is invariant with respect to reflection in this plane, we expect that the electric intensity at P will lie in this plane. Let D be the midpoint of AB and put $AP = r_1$, $BP = r_2$, and $PD = r$. Let θ be the angle PDA. The electric potential at P due to the point charge $+e$ at A is

$$\phi_1 = e/4\pi\epsilon_0 r_1.$$

The potential at P due to the charge $-e$ at B is

$$\phi_2 = -e/4\pi\epsilon_0 r_2.$$

Using the principle of superposition which we discussed earlier, we can write the total potential ϕ at P as the sum of these component potentials, i.e.

$$\phi = \frac{e}{4\pi\epsilon_0}\left(\frac{1}{r_1} - \frac{1}{r_2}\right).$$

If we use the cosine rule, we can express both r_1 and r_2 in terms of the variables r and θ. This will prove more convenient later in the problem. We have

$$\left. \begin{aligned} r_1^2 &= r^2 + d^2 - 2rd \cos\theta \\ r_2^2 &= r^2 + d^2 + 2rd \cos\theta \end{aligned} \right\}.$$

In terms of the new variables, the potential becomes

$$\phi(r,\theta) = \frac{e}{4\pi\epsilon_0} \left[\frac{1}{(r^2+d^2-2rd\cos\theta)^{1/2}} - \frac{1}{(r^2+d^2+2rd\cos\theta)^{1/2}} \right].$$

Suppose we now restrict ourselves to consider points with the distance r considerably greater than d. We can then expand the brackets in the potential, using the binomial theorem, to obtain

$$\phi(r,\theta) = \frac{e}{4\pi\epsilon_0 r} \left[\left(1 - \frac{2d\cos\theta}{r} + \frac{d^2}{r^2}\right)^{-1/2} - \left(1 + \frac{2d\cos\theta}{r} + \frac{d^2}{r^2}\right)^{-1/2} \right]$$

$$= \frac{e}{4\pi\epsilon_0 r} \left[\left(1 + \frac{d\cos\theta}{r} + \cdots\right) - \left(1 - \frac{d\cos\theta}{r} + \cdots\right) \right],$$

where we have retained terms of order (d/r). Thus to this order of approximation we have

$$\phi(r,\theta) = \frac{2ed\cos\theta}{4\pi\epsilon_0 r^2}.$$

Now let us consider the limiting case of the configuration in which the distance between the charges tends to zero, while the magnitude of each charge tends to infinity, in such a way that the product ed remains constant. Let $\mathbf{m}$ be a vector in the direction BA with magnitude $m = 2ed$. Then we can write the potential in the form

$$\phi(\mathbf{r}) = \mathbf{m}.\mathbf{r}/4\pi\epsilon_0 r^3$$

where the origin is located at the position of the charges, which in this limit coincide. This potential is the exact potential for this limiting configuration, which is known as an *electric dipole*. The higher order terms in the approximation vanish in the limit.

From this potential for an electric dipole one can derive the electric intensity, which is

$$\mathbf{E}(\mathbf{r}) = -\nabla\phi(\mathbf{r}) = \frac{1}{4\pi\epsilon_0}\left(-\frac{\mathbf{m}}{r^3} + \frac{3(\mathbf{m}.\mathbf{r})\mathbf{r}}{r^5} \right),$$

again a result of some importance in the theory of fields in media.   ◀

3.2. The Physical Significance of the Electric Potential

Let us consider a more direct physical interpretation of the electric potential describing a static field. Suppose we have a test body of charge q in an electrostatic field. Suppose the charge is moved from a point $\mathbf{r}_1$ to a second point $\mathbf{r}_2$. Let the work done against the electric intensity during this move be W, given by

$$W = -q \int_{\mathbf{r}_1}^{\mathbf{r}_2} \mathbf{E}.\mathbf{dr}. \qquad (3.11)$$

We do not need to specify the path along which the integral is to be evaluated: since $\mathbf{\nabla} \times \mathbf{E} = 0$ for a static system, the electric field is conservative, i.e. the line integral along some curve C of the tangential component of $\mathbf{E}$ is only dependent on the end-points of the curve. In terms of the electrostatic potential we can write equation (3.11) as

$$W = q \int_{\mathbf{r}_1}^{\mathbf{r}_2} \mathbf{\nabla}\phi.\mathbf{dr} = q \int_{\mathbf{r}_1}^{\mathbf{r}_2} d\phi = q[\phi(\mathbf{r}_2) - \phi(\mathbf{r}_1)]. \qquad (3.12)$$

From equation (3.12) we see that the potential difference between two points P and Q is the work done per unit charge in taking a test charge from P to Q. We often give the potential at a point a definite value by taking a point at a great distance from the system to have zero potential, but this is not necessary since the absolute value of potential has no physical significance; only the potential difference between points is measurable. We must always be sure that if we describe any physically measurable quantity as a function of the potential, then it does not depend on the absolute value of the potential at a point, but only on that part of the potential which gives rise to a potential difference between different points.

Now let us try and relate the concept of electric potential to the energy of a charged body. If we have a test charge q of mass m moving in a given electrostatic field (we ignore any fields produced by the test charge itself) with velocity $\mathbf{v}_1$ at the point $\mathbf{r}_1$ and velocity $\mathbf{v}_2$ at the point $\mathbf{r}_2$, then the loss in kinetic energy of the charge in moving from $\mathbf{r}_1$ to $\mathbf{r}_2$ will be equal to the work done against the electric intensity. If we use equation (3.12) to express this relation we get

$$\tfrac{1}{2}m(v_1^2 - v_2^2) = W = q[\phi(\mathbf{r}_2) - \phi(\mathbf{r}_1)]. \tag{3.13}$$

We can rewrite this equation in the form

$$\tfrac{1}{2}mv_1^2 + q\phi(\mathbf{r}_1) = \tfrac{1}{2}mv_2^2 + q\phi(\mathbf{r}_2), \tag{3.14}$$

and thus we see that the quantity

$$\tfrac{1}{2}mv^2 + q\phi$$

is a constant of the motion. We wish to consider this expression as the energy of the test charge, since then we would have a very useful conservation law; so we define the quantity $q\phi$ to be the potential energy of a charge q at a point where the electrostatic potential is ϕ. We must be careful however not to interpret the electric potential ϕ for a time-dependent field in the above way, since for a nonstatic field the electric intensity $\mathbf{E}$ does not form a conservative field.

The unit in which we measure the potential difference between two points is the *volt*.

3.3. LINES OF FORCE

Sometimes when discussing an electric or magnetic field, it is helpful to have some form of picture to study. Two concepts which help to form this picture are 'lines of force' and 'equipotential surfaces'. For the moment we shall restrict our considerations to the electrostatic field.

Suppose that in some region we have an electrostatic field and that we draw some surfaces on which the electric potential has constant values. For instance, in the case of a point charge these *equipotential surfaces*, as they are called, would be a set of concentric spheres with the charge as centre. If we draw a family of these surfaces so that the potentials on adjacent surfaces differ by a constant fraction of a unit, we can use these surfaces to give us information, as follows. In those places where the surfaces are close together we would expect rapid changes of the potential as we move normal to them. Since the rate of change of potential ϕ in a direction normal to such a surface is related to $\nabla\phi$ we would expect large magnitudes for the electric intensity at these places. Similarly, where the surfaces are separated by relatively larger distances, we expect much smaller values for E, the magnitude of the electric intensity.

A *line of force* is a curve in space with the tangent at each point of the curve in the direction of the electric intensity at that point. If for any region we draw a few lines of force we often get quite a good idea of the directions of the intensity distribution in that region. However, although lines of force have tangents in the direction of the electric intensity, one must not think of them as representing the path which a charged particle would follow if placed in the field. The two curves are in general quite different. As an example of lines of force, those of the field due to a point charge are straight lines through the charge. The lines of force can also be used to give us some idea of the magnitude of the electric intensity in a region. It is generally true that the closer together the lines of force are the greater the electric intensity, assuming that we only draw lines of force so that they begin and end on charges. If at any point a line of force cuts itself then we can deduce that at this point the electric intensity is zero, a point of equilibrium, since otherwise it would have to be tangential to two different curves at once.

There is a close connection between lines of force and equipotential surfaces for a static field. We know that, given a function $\phi(\mathbf{r})$, the vector $\boldsymbol{\nabla}\phi$ at any point $\mathbf{r}_0$ is normal to the surface on which $\phi(\mathbf{r}) = \phi(\mathbf{r}_0)$ which passes through the point $\mathbf{r}_0$; also at any point $\mathbf{E} = -\boldsymbol{\nabla}\phi$; thus at every point the line of force cuts the equipotential surface normally. This is obvious in the case of the point charge we have discussed above.

Now let us return to the more quantitative considerations of the electrostatic field. Since the solution of the differential equations involved in a problem of electrostatics is subject to boundary conditions on surfaces bounding the system, it is necessary to consider briefly some properties of certain materials which could form these bounding surfaces.

3.4. CONDUCTORS AND INSULATORS

In some substances there are electrons which are not tightly bound to atoms and which can move fairly freely under any applied electric field. If a substance possesses a lot of these so-called 'conduction electrons' we say that it is a good *conductor*. The more conduction electrons a substance has, the greater the current density for a given electric field. If a substance possesses few conduction electrons we say that it is an *insulator*. However, materials cannot be sharply divided into two classes, conductors and insulators: they form a continuous range from very good conductors to very bad conductors.

It is possible, for instance by using an electric battery, to maintain a constant potential difference between two points of space. We shall not discuss the mechanism of the battery which makes this possible but merely take it as a fact. This means that it is possible to have a system of conductors on which the potential is specified and constant. If we were to have a conductor with a non-uniform potential, then the conduction electrons would flow under the influence of the electric field until the potential differences were annulled. A system of conductors at prescribed potentials often forms the boundary conditions for an electrostatics problem, and we shall now consider some rather simple examples of these problems.

First let us consider examples in which we have plane symmetry.

▶ *Example*

Two infinite plane parallel conductors are maintained at constant potentials ϕ_1 and ϕ_2 volts. If the separation of the conductors is d metres, find the electric potential in volts at a point x metres from the conductor at the lower potential, when there is no charge in the region between the conductors.

▶ *Solution*

Let the potential of the two conductors be ϕ_1 and ϕ_2 volts with $\phi_1 < \phi_2$. Take a system of rectangular cartesian co-ordinates (x, y, z) with the x-axis normal to the planes of the conductors. Let the conductors occupy the planes $x = 0$ and $x = d$. Let $\phi(x)$ be the potential at a point distance x metres from the plate at potential ϕ_1. By the symmetry of the system, $\phi(x)$ depends only on the x-co-ordinate and not on y or z. The boundary conditions of this problem are thus

$$\phi(0) = \phi_1, \qquad \phi(d) = \phi_2. \tag{a}$$

For the region between the conductors we have the electric potential satisfying the Laplace* equation

$$\nabla^2 \phi = 0,$$

---

* P. S. Marquis de Laplace (1749–1827), French mathematician and astronomer, renowned for his work on gravitation.

which in this one-dimensional problem reduces to

$$\frac{d^2\phi}{dx^2} = 0.$$

Our mathematical problem is to solve this equation subject to the above boundary conditions on ϕ.

Direct integration of the above equation twice gives

$$\phi(x) = Ax + B,$$

where A and B are constants to be determined from the boundary conditions (a).

Using the boundary conditions (a), we obtain

$$\phi_1 = A.0 + B,$$
$$\phi_2 = A.d + B,$$

and hence

$$A = (\phi_2 - \phi_1)/d \quad \text{and} \quad B = \phi_1.$$

Thus the general solution of the problem is

$$\phi(x) = [(\phi_2 - \phi_1)/d]x + \phi_1 \text{ volts.} \qquad \blacktriangleleft$$

Let us now consider a more complex problem, which does not have the plane symmetry which made the last problem so easy to solve.

▸ *Example*

Five faces of a cube are made of conducting material and maintained at zero potential. The remaining face is missing, giving us an open box. If we are given the electric potential at all points of the open face of the cube, determine the potential at all points inside the cube. All charges may be assumed to reside on the conductor.

▸ *Solution*

Choose a system of rectangular Cartesian co-ordinates so that the conducting faces of the cube are $x=0$, $x=a$, $y=0$, $y=a$, $z=0$. The open face is then $z=a$. The problem we have to solve is to obtain a solution of Laplace's equation, since the region is free from charge, which is zero on the conductor and agrees with the given potential at the open face; i.e. find a potential $\phi(x, y, z)$ such that

$$\frac{\partial^2\phi}{\partial x^2} + \frac{\partial^2\phi}{\partial y^2} + \frac{\partial^2\phi}{\partial z^2} = 0 \qquad \text{(i)}$$

and which satisfies the boundary conditions

$$\phi = 0 \quad \text{on} \quad x = 0, x = a, y = 0, y = a, z = 0; \qquad \text{(ii)}$$

$$\phi = f(x, y) \quad \text{on} \quad z = a, \qquad \text{(iii)}$$

where $f(x, y)$ is a given function which vanishes at the edges of the face in the plane $z = a$, this being necessary for the conditions to be consistent.

A solution of equation (i) which satisfies the conditions (ii) is readily found by the method of separation of variables, this being

$$\phi(x, y, z) = \sum_m \sum_n a_{mn} \sin(m\pi x/a) \sin(n\pi y/a) \sinh(b_{mn}z), \qquad \text{(a)}$$

where m and n are integers and

$$b_{mn} = \pi(m^2 + n^2)^{1/2}/a. \qquad \text{(b)}$$

Now we must see whether it is possible to make this function satisfy the boundary condition (iii).

We can express the function $f(x, y)$ as a double Fourier series as follows:

$$f(x, y) = \sum_m \sum_n c_{mn} \sin(m\pi x/a) \sin(n\pi y/a), \qquad \text{(c)}$$

where the coefficients are given by the Fourier expressions

$$c_{mn} = \frac{4}{a^2} \int_0^a \int_0^a f(x, y) \sin(m\pi x/a) \sin(n\pi y/a) \, dx \, dy. \qquad \text{(d)}$$

Expression (c) shows the vanishing of f at the open-face edges. The function (a) will satisfy the boundary condition (iii) if we put

$$a_{mn} = c_{mn} \operatorname{cosech}(b_{mn}a), \qquad \text{(e)}$$

since then the series (a) and (c) are identical on $z = a$, as required. Inserting the coefficients a_{mn} from equation (e) into equation (a), we see that the solution we seek is

$$\phi(x, y, z) = \sum_m \sum_n c_{mn} \operatorname{cosech}(b_{mn}a) \sin(m\pi x/a) \sin(n\pi y/a) \sinh(b_{mn}z)$$

with the coefficients c_{mn} and b_{mn} given by equations (d) and (b) respectively.   ◀

We have considered two examples involving systems with plane surfaces; now let us turn to one which involves spherical surfaces.

▶ *Example*

An uncharged conducting sphere of radius a is placed in an electric field which is uniform and of intensity $\mathbf{E}_0$ at large distances from the sphere. Find the potential at all points outside the sphere.

▶ *Solution*

Take the origin at the centre of the sphere and use spherical polar co-ordinates with the polar axis in the direction of $\mathbf{E}_0$. We seek a potential $\phi(r, \theta, \psi)$, such that if $\mathbf{E}$ is the corresponding electric intensity we have

$$\nabla^2 \phi = 0; \tag{1}$$

$$\phi = \text{constant on } r = a; \tag{2}$$

$$\mathbf{E} \to \mathbf{E}_0 \quad \text{as} \quad r \to \infty. \tag{3}$$

The boundary condition (2) is a result of the properties of conductors we referred to earlier: they must be equipotential surfaces in a static electric field. The value of the constant in condition (2) is immaterial since to any potential we can add constants without altering the corresponding electric intensity, which is the observable quantity.

In spherical polar co-ordinates we would expect from the symmetry of the system about the polar axis that the electric potential ϕ would not depend on the azimuthal angle ψ, which describes position round this axis, but only on the variables r and polar angle θ. A solution of equation (1) with this property is

$$\phi(r, \theta) = \sum_{n=1}^{\infty} (A_n r^n + B_n r^{-n-1}) P_n(\cos \theta) \tag{4}$$

where the P_n are the Legendre* polynomials. This form of solution is obtained by the separation of variables method. The constants A_n and B_n are to be obtained from the boundary

* A. M. Legendre (1752–1833), French mathematician whose main work was on elliptic integrals.

conditions of the problem. Since condition (3), which can be written in the equivalent form

$$\phi(r,\,\theta) \to -E_0 r \cos\theta \quad \text{as} \quad r \to \infty, \tag{5}$$

only contains the first power of $\cos\theta$, we shall try a solution of the more restricted type

$$\phi(r,\,\theta) = (Ar + Br^{-2}) \cos\theta. \tag{6}$$

Remembering that $P_1(\cos\theta) = \cos\theta$ we see that this is merely a special case of (4).

If we subject the solution (6) to the condition (5), we see that $A = -E_0$. Condition (2), provided we take the constant to be zero, leads to

$$Aa + Ba^{-2} = 0.$$

Thus $B = -Aa^3 = E_0 a^3$ and the solution of the problem is

$$\phi(r,\,\theta) = -E_0\left(1 - \frac{a^3}{r^3}\right)r\cos\theta.$$

In this form the asymptotic behaviour as $r \to \infty$ is easily seen. ◀

For a discussion of the uniqueness of the solutions to the problems we have considered, see Problem 5 at the end of this chapter.

We shall now consider an example which will show the behaviour of the electric intensity at a point near the surface of a conductor and the forces which will act on the surface of a conductor due to charge residing there.

▶ *Example*

Consider a conductor (Fig. 4) having a surface S on which the charge density, which in general will be a function of position,

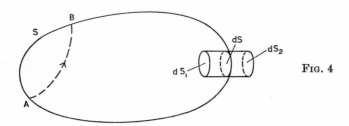

FIG. 4

will be denoted by σ. We shall show that at any point outside the conductor, but very close to it, the electric intensity is normal to the surface and has a magnitude of σ/ϵ_0. We shall also show that due to this electric intensity the charge on S is subject to an outward normal force such that the force per unit area is $\sigma^2/2\epsilon_0$.

▶ *Solution*

The first step in this problem is to prove an important result about conductors in fields. This result is that under static conditions one cannot have an electric field inside a conductor. Since S is a conductor which has free electrons in it, which would move in response to any tangential component of electric intensity, the static equilibrium situation must be one in which the electric intensity is everywhere normal to the surface S. Thus the surface S will be an equipotential surface. Let us suppose that there is an electric field inside the conducting surface S. A line of force can only begin and end on charges, since if this was not so, then a small surface which surrounded the end-point would give a flux which violated Gauss' Law * given in equation (3.9). We see that any lines of force which exist inside S must begin and end on S. (If the conductor is hollow then obviously the charge resides in the surface S; and even if the conductor is solid, we show in Problem 1 of this chapter that for a static situation there is no charge inside S.) Let such a line of force begin at A and end at B. As one proceeds from A to B along the line of force, the line integral of the electric intensity is nonzero and hence the potential of B is not that of A, which is a contradiction of the fact that S is an equipotential surface. Hence the conclusion we can draw is that the electric intensity is zero everywhere inside S: we say that the region is 'field-free'.

This result could have been established in a more mathematical way by using properties of harmonic functions and showing that the electric potential is constant inside a conducting surface.

Now consider a small element dS of the surface S. Suppose we erect on this element dS a small cylinder with generators normal to S and end faces dS_1 and dS_2 parallel to dS, as shown in

---

* C. F. J. Gauss (1777–1855), a first-class physicist, is believed by many to have been the greatest of all mathematicians.

Fig. 4. The total charge contained inside this cylinder, by definition of the surface density of charge σ on S, is given by $\sigma\,dS$.
Now consider the flux of the electric intensity out of the surface
of this cylinder. Over the face dS_1 the intensity is zero, a fact
we have just proved. Since the electric intensity at points just
outside S is normal to S, there is no flux across the curved
surface of the cylinder if the length of the generators is small.
Over dS_2 the intensity is normal, and if we assume the face to be
small enough we can take the intensity to be uniform. Thus the
total flux out of the cylinder is $E\,dS_2$.

If we now use the result (3.9) we can see that we must have

$$E\,dS_2 = \sigma\,dS/\epsilon_0.$$

If we now let dS_2 approach dS and consider the limiting case we
obtain

$$E = \sigma/\epsilon_0,$$

which is the desired result.

Before we find the force on an element of the surface S let us
consider the problem of an infinite plane conducting sheet,
which contains a result we shall need.

Suppose that we have an infinite plane conducting sheet with
a uniform charge density σ per unit area. If we use the argument
with the small cylinder that we used above and exploit the
symmetry of the system, remembering of course that we do not
have a closed conducting surface but an open one, then we can
show that the electric intensity at any point is $\sigma/2\epsilon_0$ in a direction
normal to the conductor and away from it.

Now let us return to the problem of the closed conductor S.
If we are at a point P very close to dS, but outside S, the intensity at P due to dS is that due to an infinite plane with charge
density σ. This is because any surface looks like an infinite
plane if you are close enough to it. Hence the intensity at P
due to dS is along the outward normal and of magnitude $\sigma/2\epsilon_0$,
a result we discussed in the previous paragraph. If we are at a
point Q very close to dS, but inside S, the intensity there due to
dS is along the inward normal and of magnitude $\sigma/2\epsilon_0$.

Now consider the electric intensity due to the rest of the
surface S. At Q, since there is no intensity inside a conductor,
it must just balance that produced by dS. Hence it must be

along the outward normal with magnitude $\sigma/2\epsilon_0$. As we cross dS the contribution to the intensity from the rest of S must be continuous so that at P it is the same as at Q. Thus at P we have an intensity $\sigma/2\epsilon_0$ along the outward normal due to the charge on dS, and an intensity $\sigma/2\epsilon_0$ along the outward normal due to the charge on the remainder of S. The total intensity at P is thus σ/ϵ_0 along the outward normal to S. This checks our previous result by a different method.

The force on the element dS is the product of the charge on the element and the electric intensity due to the rest of the surface. This gives a force along the outward normal equal to $\sigma\, dS\, \sigma/2\epsilon_0$ which leads to a force density of $\sigma^2/2\epsilon_0$ as we set out to prove. ◀

We close this chapter with an example which introduces the idea of the mutual potential energy which a system of charged conductors possesses by virtue of the fact that, in general, work must be done to bring a system of charged conductors into a given configuration.

▶ *Example*

Show that if we have a system of conductors with charges e_i and potentials ϕ_i, then their mutual potential energy due to the forces acting between them is W, where

$$W = \tfrac{1}{2} \sum_i e_i \phi_i.$$

▶ *Solution*

Consider a system of conductors which at a time t carry a fixed fraction f of the final charge on each. By the principle of superposition the potentials of the conductors will also be the same fraction of their final potentials. The work which must be expended to increase the fractional charges by an amount df, again the same for all conductors, will be, assuming the potential to vanish at large distances from the system,

$$dW = \sum_i (f\phi_i)(df e_i)$$

where e_i and ϕ_i are the final charges and potentials on the conductors. If we now sum over all such elemental increments and pass to the limit of infinitely small increments, we obtain

$$W = \sum_i (e_i \phi_i) \int_0^1 f\, df = \tfrac{1}{2} \sum_i e_i \phi_i$$

as required. The unit of energy is the *joule*,* which is the work required to raise one coulomb of charge through a potential difference of one volt. ◀

PROBLEMS FOR CHAPTER 3

(1) Suppose that we have a uniform distribution of charge which, in cylindrical polar co-ordinates, lies along the z-axis. Show that the electric intensity is radially outward and of magnitude $q/2\pi\epsilon_0 r$, where q is the charge per unit length of the z axis. Show also that the electric potential is of the form $(C - q \log r)/2\pi\epsilon_0$, where C is a constant.

(2) Given that a region of space $0 \leqslant x \leqslant a$, $0 \leqslant y \leqslant b$, $0 \leqslant z \leqslant c$ is free from charge, and that the electric potential ϕ has values

$$\phi = 0 \quad \text{on} \quad x = 0, x = a, y = 0, y = b, z = 0,$$
$$= xy(x-a)(y-b) \quad \text{on} \quad z = c,$$

find the value of ϕ at all points inside the region.

(3) Two coaxial metal cylinders of radii a and b, with $a > b$, are maintained at potentials V_1 and V_2 respectively. Determine the electric intensity at a point between the cylinders.

(4) Show that the force exerted on an electric dipole of moment **m** by an electric field **E** is $(\mathbf{m} . \nabla)\mathbf{E}/4\pi\epsilon_0$. (Hence an electric dipole in a uniform field experiences no resultant force.)

(5) Show that if we have a system of finite conductors on each of which is prescribed either the potential or the total charge, then the potential for the region outside the conductors which is compatible with these conditions is unique.

(6) Suppose we have a pair of concentric spherical conductors carrying equal and opposite charges. Determine the potential difference between them in terms of the charge on each and their radii.

---

* J. P. Joule (1818–1889), English physicist, the pioneer of work on the interconvertibility of various forms of energy.

CHAPTER 4

THE MAGNETOSTATIC FIELD

In this chapter we shall consider situations in which we have only a static magnetic field. As in the previous chapter, we shall find that the condition that the field be static leads to considerable simplification of the general results obtained in Chapter 2.

4.1. THE STATIC MAGNETIC FIELD

Suppose that we have a system in which all the motions of charge are steady, i.e. the current density $\mathbf{j}$ is independent of time. The current density in a system is normally caused by a flow of electrons, the positive charges residing on the much less mobile atomic nuclei. However, although the heavy nuclei do not normally move to produce a current, they do provide a background of positive charges which neutralise the electric field of the electrons, at least on a macroscopic scale. Thus we can have an electrically neutral system with a current flowing, due to the influence of some externally applied electric field, which produces a magnetic field but not an electric field. In this chapter we shall study such systems.

In the steady conditions considered the field equations (1.5) and (1.7) reduce to

$$\mathbf{\nabla} \times \mathbf{B} = \mu_0 \mathbf{j}, \tag{4.1}$$

$$\mathbf{\nabla} . \mathbf{B} = 0. \tag{4.2}$$

The vector potential for such a field, defined by

$$\mathbf{B} = \mathbf{\nabla} \times \mathbf{A}, \tag{4.3}$$

satisfies the equation

$$\nabla^2 \mathbf{A} = -\mu_0 \mathbf{j}, \tag{4.4}$$

45

which is the restricted case of (2.19). The solution of this equation for the steady state is

$$\mathbf{A}(\mathbf{r}) = \frac{\mu_0}{4\pi} \int \int \int \frac{\mathbf{j}(\mathbf{r}')}{|\mathbf{r} - \mathbf{r}'|} \, d\tau, \qquad (4.5)$$

where the volume integral is taken over the whole region occupied by the current distribution.

Very often in practical applications of electromagnetic theory we are interested in calculating the magnetic field produced by a current flowing in a conducting wire, and we shall now show how this can be done using equation (4.5).

Suppose we have a thin wire in the shape of a curve C. Let the wire carry a current J. If we consider equation (4.5) we can take for the volume element $d\tau$ a small section of the wire, represented vectorially by $\mathbf{ds}'$ in the direction of the current flow, and write $\mathbf{j}(\mathbf{r}') \, d\tau$ in the form $J \, \mathbf{ds}'$. The equation (4.5) for the magnetic vector potential due to a current J flowing in the wire C becomes

$$\mathbf{A}(\mathbf{r}) = \frac{\mu_0 J}{4\pi} \int_C \frac{\mathbf{ds}'}{|\mathbf{r} - \mathbf{r}'|} \qquad (4.6)$$

where $\mathbf{r}'$ is the position of the line element $\mathbf{ds}'$ of C. The sense of description of the curve C is the direction of the current flow.

From equation (4.6) for the magnetic vector potential we can calculate the magnetic intensity for a current flowing in a wire. We have

$$\mathbf{B}(\mathbf{r}) = \mathbf{\nabla} \times \mathbf{A}(\mathbf{r}) = \frac{\mu_0 J}{4\pi} \mathbf{\nabla} \times \int_C \frac{\mathbf{ds}'}{|\mathbf{r} - \mathbf{r}'|}.$$

If we assume that we can differentiate under the integral sign, this gives

$$\mathbf{B}(\mathbf{r}) = \frac{\mu_0 J}{4\pi} \int_C \mathbf{\nabla} \times \left(\frac{\mathbf{ds}'}{|\mathbf{r} - \mathbf{r}'|} \right). \qquad (4.7)$$

We know that

$$\mathbf{\nabla} \times (\phi \mathbf{a}) = \phi \mathbf{\nabla} \times \mathbf{a} - \mathbf{a} \times (\mathbf{\nabla} \phi)$$

and so we can write equation (4.7) as

$$\mathbf{B}(\mathbf{r}) = -\frac{\mu_0 J}{4\pi} \int_C \mathbf{ds}' \times \mathbf{\nabla} \left(\frac{1}{|\mathbf{r} - \mathbf{r}'|} \right) = \frac{\mu_0 J}{4\pi} \int_C \frac{\mathbf{ds}' \times (\mathbf{r} - \mathbf{r}')}{|\mathbf{r} - \mathbf{r}'|^3}. \qquad (4.8)$$

This result (4.8) for the magnetic intensity for a current flowing in a wire is known as the *Biot–Savart** Law. Thus to calculate such an intensity we can either use equation (4.8) directly or we can calculate the magnetic vector potential using equation (4.6) and then obtain the intensity from this using $\mathbf{B} = \nabla \times \mathbf{A}$, whichever is the more convient for the problem in hand.

When considering magnetic fields it is sometimes more convenient, as with electric fields, to use an integral form of the field equations. This is often so if the system has any symmetry. We shall now consider this integral form of the field equation, and then illustrate each method of dealing with a field by a simple example.

Consider an open surface S bounded by a closed curve C. Integrate both sides of equation (4.1) over this surface, obtaining

$$\iint_S (\nabla \times \mathbf{B}) . \, d\mathbf{S} = \mu_0 \iint_S \mathbf{j} . \, d\mathbf{S}. \tag{4.9}$$

The integral on the right-hand side of this equation is merely the total current crossing the surface S. If we use Stokes' Theorem we can write the equation as

$$\int_C \mathbf{B} . \, d\mathbf{s} = \mu_0 J \tag{4.10}$$

with J as the total current crossing the surface S. This is the integral form of the field equations we referred to above.

▶ *Example*

Determine the field near a long straight wire carrying a current J.

▶ *Solution*

If we stay away from the ends of the wire and near to the wire we can assume the wire to be infinite for the purposes of calculation. We would expect from the symmetry of the system that the magnetic intensity in this region would be invariant under a translation parallel to the wire, which means that if we use cylindrical polar co-ordinates (r, θ, z) with the wire as z-axis we would expect the magnetic intensity $\mathbf{B}$ to be independent of z. From the rotational symmetry about the z-axis, we would expect that B would also be independent of θ, since

---

* J. B. Biot (1774–1862) French physicist. His main work was concerned with the polarisation of light. F. Savart (1791–1841), French physicist.

B is an observable. It would be possible to have the magnitude of the magnetic vector potential **A** as a function of θ, but **A** is not an observable and our symmetry considerations would not apply. The differentiation of **A** to give **B** would remove the asymmetry. Hence B is a function of the distance r of the field point from the wire only. From equation (4.8) we see that the direction of **B** is tangential to a circle with its centre on the wire and its plane normal to the wire. If we use these considerations in equation (4.10) we obtain

$$2\pi r B(r) = \mu_0 J;$$

hence

$$B(r) = \mu_0 J / 2\pi r.$$

We can write the magnetic intensity in cylindrical polar coordinates, if J is in the positive z direction as

$$\mathbf{B(r)} = (0, \mu_0 J / 2\pi r, 0),$$

and we have solved our initial problem.   ◀

We now turn our attention to a problem which introduces a very important concept into magnetic theory, the magnetic dipole. This is useful in the theory of magnetic fields in matter in the same way that the electric dipole forms a starting-point for the discussion of electric fields in matter.

▶ *Example*

Determine the magnetic vector potential and magnetic intensity of a plane current-carrying wire loop at points whose distances from the loop are large compared with its maximum linear dimension.

▶ *Solution*

Let C be the plane current loop and J the current flowing in it. Let S be a plane surface bounded by C. Consider the magnetic potential at a point **r**. From equation (4.6) we have

$$\mathbf{A(r)} = \frac{\mu_0 J}{4\pi} \int_C \frac{\mathbf{ds'}}{|\mathbf{r-r'}|},$$

with our usual notation.

We know that for a scalar function $\phi(\mathbf{r}')$ we have the result

$$\int_C \phi \, d\mathbf{s}' = \int\int_S d\mathbf{S}' \times \nabla'\phi;$$

hence we can write the vector potential in the form

$$\mathbf{A}(\mathbf{r}) = \frac{\mu_0 J}{4\pi} \int\int_S d\mathbf{S}' \times \nabla'\left(\frac{1}{|\mathbf{r}-\mathbf{r}'|}\right) = \frac{\mu_0 J}{4\pi} \int\int_S \frac{d\mathbf{S}' \times (\mathbf{r}-\mathbf{r}')}{|\mathbf{r}-\mathbf{r}'|^3}.$$

If we are at distances from the loop which are large compared to the maximum linear dimension of the loop, we can write

$$\mathbf{A}(\mathbf{r}) = \frac{\mu_0}{4\pi} \frac{[\mathbf{m} \times (\mathbf{r}-\mathbf{r}')]}{|\mathbf{r}-\mathbf{r}'|^3}, \tag{a}$$

where $\mathbf{m} = J \times$ (vector area of S) and $\mathbf{r}'$ is the position vector of some point in S.

We refer to field configurations of the form (a) as dipole fields and we call $\mathbf{m}$ the *magnetic moment* of the dipole or arrangement producing the field, in this case a current loop.

If we derive the magnetic intensity from (a) for the dipole field we obtain an expression which is similar to the electric field due to an electric dipole.

$$\mathbf{B}(\mathbf{r}) = \frac{\mu_0}{4\pi}\left(\frac{-\mathbf{m}}{|\mathbf{r}-\mathbf{r}'|^3} + \frac{3[\mathbf{m}\cdot(\mathbf{r}-\mathbf{r}')](\mathbf{r}-\mathbf{r}')}{|\mathbf{r}-\mathbf{r}'|^5}\right). \tag{b}$$

In some alternative approaches to the theory of the magnetic field the equations (a) and (b) for magnetic dipoles play a fundamental part.

It is well known that certain substances are capable of producing magnetic fields without our causing any macroscopic current to flow in them. The magnetic field in such cases is caused by currents flowing on the microscopic scale. It is found that if we take a small sample of one of these so called 'permanently magnetised' substances, we obtain the dipole fields we have studied in the above example, provided we observe the restriction on distance from the sample. Hence in a way magnetic dipoles exist in nature. We may consider the electrons which circulate round the nucleus in an atom as being small current loops giving rise to some of the magnetic properties of matter.

Now let us consider a problem of a more practical nature.

▶ *Example*

Consider a wire (Fig. 5) wound closely on to a cylinder of radius R. The wire is wound at N turns per metre for a distance $2R$. The wire carries a current J amperes. Find the magnetic intensity at the centre of the coil.

▶ *Solution*

Let O be the centre of the coil. Consider a strip of coil between two planes normal to the axis of the coil at distances x and $x+dx$ from O. The total current flowing around the strip is $JN\,dx$. Consider the intensity $\mathbf{dB}$ at the centre of the coil due to this strip.

$$\mathbf{dB} = \frac{\mu_0}{4\pi} \int_C (JN\,dx) \frac{\mathbf{ds}\times\mathbf{r}}{r^3} \tag{1}$$

where $\mathbf{r}$ is the vector from the strip $\mathbf{ds}$ to O. From the symmetry of the system the intensity at O must be parallel to the axis of the coil, which we choose as the x-axis. If we now sum the contributions from all such strips, we obtain for the intensity at O:

$$B = \frac{\mu_0 NJ}{4\pi} \int_{-a}^{a} dx \int_0^{2\pi} \frac{a\,d\theta . r}{r^3} \cdot \frac{a}{r}$$

$$= \frac{\mu_0 NJa^2}{2} \int_{-a}^{a} \frac{dx}{(a^2+x^2)^{3/2}} = \frac{\mu_0 NJ}{2^{1/2}}. \quad ◀$$

▶ *Example*

Determine the magnetic intensity at the centre of a circular wire of radius R m carrying a current J amperes (Fig. 6).

▶ *Solution*

The field is parallel to the unit normal $\mathbf{n}$ of the plane of the wire C by symmetry. From the Biot–Savart law we obtain

$$\mathbf{B(r)} = \frac{\mu_0}{4\pi} J \int_C \frac{\mathbf{ds'}\times(\mathbf{r}-\mathbf{r'})}{|\mathbf{r}-\mathbf{r'}|^3} \; \text{Wb/m}^2.$$

$$= \mathbf{n}\,\frac{J\mu_0}{4\pi} \int_0^{2\pi} \frac{R\,d\theta . R}{R^3} = \frac{\mu_0 J}{2R}\,\mathbf{n} \; \text{Wb/m}^2. \quad ◀$$

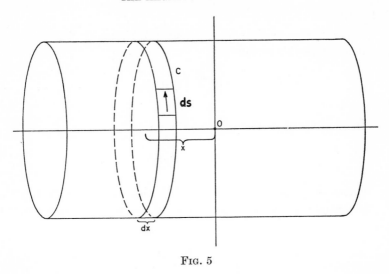

FIG. 5

Very often in practical problems one has two or more circuits carrying currents which interact with each other. One may for instance be interested in the electromotive force produced in one circuit by changing the current flowing in another circuit, which will of course produce a change in the flux linking the first circuit. In this context it is convenient to define concepts of 'self-inductance' for a circuit and 'mutual inductance' for pairs of circuits. We consider an example.

▶ *Example*

The *self-inductance* of a circuit C is defined to be the flux of magnetic intensity which links the circuit when unit current flows in it. Obtain an expression for this quantity.

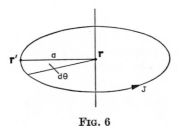

FIG. 6

▶ *Solution*

Let S be a surface bounded by the circuit C. The magnetic flux across this surface is

$$\Phi = \int\int_S \mathbf{B}.\mathbf{dS}. \tag{1}$$

If we use Stokes' Theorem we can write

$$\Phi = \int_C \mathbf{A}.\mathbf{ds}, \tag{2}$$

where as usual $\mathbf{A}$ is the magnetic vector potential. If $\mathbf{A}$ is the potential due to unit current flowing in C we have shown that

$$\mathbf{A(r)} = \frac{\mu_0}{4\pi} \int_C \frac{\mathbf{ds'}}{|\mathbf{r}-\mathbf{r'}|}, \tag{3}$$

where $\mathbf{ds'}$ is the element of C located at the point $\mathbf{r'}$. From equations (3) and (2) we see that

$$\Phi = \frac{\mu_0}{4\pi} \int_C\int_C \frac{\mathbf{ds(r)}.\mathbf{ds'(r')}}{|\mathbf{r}-\mathbf{r'}|}, \tag{4}$$

which by definition is the self-inductance of the circuit C. ◀

▶ *Example*

The *mutual inductance* of two circuits C and C' is defined to be the flux of the magnetic intensity through one due to unit current in the other. Determine the mutual inductance of two square circuits, each of side a, placed in parallel planes at a distance a apart so as to form boundaries of the opposite faces of a cube (Fig. 7).

▶ *Solution*

By an argument which is almost identical to that used in obtaining the self-inductance of a circuit, we can show that the mutual inductance of C and C' is M, where

$$M = \frac{\mu_0}{4\pi} \int_C\int_{C'} \frac{\mathbf{ds}.\mathbf{ds'}}{|\mathbf{r}-\mathbf{r'}|},$$

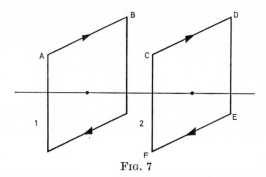

FIG. 7

where $\mathbf{ds}$ is an element of C at the point $\mathbf{r}$ and $\mathbf{ds}'$ is an element of C' at the point $\mathbf{r}'$.

Now let us consider the particular circuits we are given. From the fact that the expression for the mutual induction contains the scalar product of the two elements of the circuits we see that no contribution is made by pairs of sides which are mutually perpendicular. We also see by symmetry that we can group the remaining pairs of sides. Referring to the figure, we obtain

$$M = \frac{\mu_0}{\pi} \left(\int_{AB} \int_{CD} \frac{\mathbf{ds} . \mathbf{ds}'}{|\mathbf{r} - \mathbf{r}'|} + \int_{AB} \int_{EF} \frac{\mathbf{ds} . \mathbf{ds}'}{|\mathbf{r} - \mathbf{r}'|} \right).$$

Let x_1 and x_2 be the distances of elements from the centre points of their corresponding sides. Then we have

$$M = \frac{\mu_0}{\pi} \int_{-a/2}^{a/2} dx_1 \int_{-a/2}^{a/2} \frac{dx_2}{[(x_1 - x_2)^2 + a^2]^{1/2}}$$
$$- \frac{\mu_0}{\pi} \int_{-a/2}^{a/2} dx_1 \int_{-a/2}^{a/2} \frac{dx_2}{[(x_1 - x_2)^2 + 2a^2]^{1/2}}.$$

After some rather tedious substitutions, this reduces to

$$M = \frac{2\mu_0}{\pi} \left[\log \left(\frac{2 + 2^{1/2}}{1 + 3^{1/2}} \right) + 3^{1/2} - 2^{3/2} + 1 \right]$$

which is the result we set out to obtain. ◀

In this chapter we have obtained some of the more interesting and useful results of the theory of the magnetostatic field. The following problems also contain important results and should be studied carefully. However, we shall close this chapter with a final example on the dipole which contains an important result for the force on such a system.

▶ *Example*

Show that the force exerted by a static magnetic field $\mathbf{B}(\mathbf{r})$ on a magnetic dipole of moment $\mathbf{m}$ at the point $\mathbf{r}$ is $(\mathbf{m}.\nabla)\mathbf{B}(\mathbf{r})$.

▶ *Solution*

We shall represent the dipole by a small plane current carrying loop C. Let J be the current in the loop. The total force on the loop C due to the magnetic field is

$$\mathbf{F} = J \int_C \mathbf{ds} \times \mathbf{B}.$$

We must now transform this into a more convenient form. Let $\mathbf{a}$ be a constant vector, then

$$\mathbf{a}.\int_C \mathbf{B} \times \mathbf{ds} = \int_C \mathbf{a}.(\mathbf{B} \times \mathbf{ds}) = \int_C (\mathbf{a} \times \mathbf{B}).\mathbf{ds}.$$

Stokes' theorem applied to a smooth surface S bounded by C gives us

$$\mathbf{a}.\int_C \mathbf{B} \times \mathbf{ds} = \int\int_S [\nabla \times (\mathbf{a} \times \mathbf{B})].\mathbf{dS}.$$

But

$$\nabla \times (\mathbf{a} \times \mathbf{B}) = \mathbf{a}(\nabla.\mathbf{B}) - \mathbf{B}(\nabla.\mathbf{a}) + (\mathbf{B}.\nabla)\mathbf{a} - (\mathbf{a}.\nabla)\mathbf{B}.$$

The first term on the right-hand side is zero from the field equations, the second and third terms are zero since $\mathbf{a}$ is a constant vector. Thus we have

$$\mathbf{a}.\int_C \mathbf{ds} \times \mathbf{B} = \int\int_S [(\mathbf{a}.\nabla)\mathbf{B}].\mathbf{dS}.$$

We can now write the right-hand side of this equation as

$$-\int\int_S \mathbf{a}.\nabla(\mathbf{B}.\mathbf{dS})$$

provided that we remember that $d\mathbf{S}$ is not a vector field and that the gradient operator only acts on $\mathbf{B}$. Hence we have

$$\mathbf{a}.\int_C d\mathbf{s} \times \mathbf{B} = \mathbf{a}.\int\int_S \nabla(\mathbf{B}.d\mathbf{S}).$$

Now

$$\nabla(\mathbf{B}.d\mathbf{S}) = d\mathbf{S} \times (\nabla \times \mathbf{B}) + (d\mathbf{S}.\nabla)\mathbf{B} = (d\mathbf{S}.\nabla)\mathbf{B}$$

since the external magnetic field $\mathbf{B}$ has zero curl, since it is static and not produced by current at the dipole.

Thus we have from the two preceding equations

$$\mathbf{a}.\int_C d\mathbf{s} \times \mathbf{B} = \mathbf{a}.\int\int_S (d\mathbf{S}.\nabla)\mathbf{B}$$

and since $\mathbf{a}$ is an arbitrary vector this implies that

$$\int_C d\mathbf{s} \times \mathbf{B} = \int\int_S (d\mathbf{S}.\nabla)\mathbf{B}.$$

The force on the current loop can now be written as

$$\mathbf{F} = J \int\int_S (d\mathbf{S}.\nabla)\mathbf{B}.$$

Now let the surface S be plane and small. We can then assume the integrand is constant over the surface and write

$$\mathbf{F} = (\mathbf{m}.\nabla)\mathbf{B}(\mathbf{r})$$

where $\mathbf{m}$ is the magnetic moment of the small loop, defined by

$$\mathbf{m} = J\, d\mathbf{S}. \qquad\qquad ◀$$

PROBLEMS FOR CHAPTER 4

(1) Show that the couple exerted on a magnetic dipole of moment $\mathbf{m}$ by a magnetic field $\mathbf{B}$ is $\mathbf{m} \times \mathbf{B}$.

(2) A permanent magnetic dipole of moment $\mathbf{m}$ is suspended at a point $\mathbf{r}$ in a magnetic field so that it can rotate about an axis perpendicular to $\mathbf{m}$. The direction of the magnetic intensity $\mathbf{B}$ is parallel to $\mathbf{m}$. If the dipole is rotated through a small angle from its position of equilibrium and released, show that it will oscillate with a period

$T = 2\pi[I/mB(\mathbf{r})]^{1/2}$, where I is the moment of inertia of the dipole about the axis of rotation.

[This arrangement is the basis of the *oscillating magnetometer*, an instrument for comparing magnetic intensities. If T_1 and T_2 are the periods of oscillations at points $\mathbf{r}_1$ and $\mathbf{r}_2$ then $T_1^2 : T_2^2 :: B(\mathbf{r}_2) : B(\mathbf{r}_1)$.]

(3) Determine the force $\mathbf{F}$ and couple $\mathbf{M}$ which a dipole of moment $\mathbf{m}$ at the point $\mathbf{r}$ exerts on a second dipole of moment $\mathbf{m}'$ at the point $\mathbf{r}'$. Show that the force is equal and opposite to the force exerted by $\mathbf{m}'$ on $\mathbf{m}$.

(4) Two small magnetic needles of moments $\mathbf{m}$ and $\mathbf{m}'$ have an angle α between their directions and are rigidly connected together. Show that when suspended freely in a uniform magnetic field $\mathbf{B}$ their directions will be at angles θ and θ' with the field, where

$$(\sin \theta)/m = (\sin \theta')/m' = (\sin \alpha)/(m^2 + m'^2 + 2mm' \cos \alpha)^{1/2}.$$

(5) A small magnet ACB, free to turn about its centre C, is acted on by a small fixed magnet PQ. Show that in equilibrium the axis of ACB lies in the plane PQC and that if θ and θ' are the angles which AB and PQ make with the line joining their centres then

$$\tan \theta' = -\tfrac{1}{2} \tan \theta.$$

CHAPTER 5

ELECTROMAGNETIC WAVES

5.1. THE DISPLACEMENT CURRENT

One of the great unifications of science took place in the nineteenth century when Clerk Maxwell predicted on theoretical grounds that light was a form of electromagnetic radiation and this conjecture was later verified experimentally by Hertz.*

It had been realised that the equation of Ampère relating magnetic intensity to current, i.e.

$$\mathbf{\nabla} \times \mathbf{B} = \mu_0 \mathbf{j}, \qquad (5.1)$$

was true only for a steady state. Hence Maxwell was faced with the problem of replacing equation (5.1) by an equation which was to be valid for time-dependent fields, but which would reduce to equation (5.1) in the steady state, where the validity of equation (5.1) was known from experiment. Maxwell solved the problem by adding to the right-hand side of equation (5.1) a term which was a time derivative, and which therefore vanished in the steady state. He wrote for the field equation *in vacuo*

$$\mathbf{\nabla} \times \mathbf{B} = \mu_0\left(\mathbf{j} + \epsilon_0 \frac{\partial \mathbf{E}}{\partial t}\right). \qquad (5.2)$$

The added term is referred to as the *displacement current*. It must not be thought of as a flow of charge, since it can exist in regions free from charge, but it can produce magnetic fields in the same manner as a conduction current. Of course we have used this complete field equation (5.2) from the beginning in our discussions of the electromagnetic field.

We shall see in the following paragraphs that the presence of the

---

* H. R. Hertz (1857–1894), the German physicist who first detected electromagnetic waves.

displacement current in the field equations is of paramount importance and allows the possibility of having an electromagnetic field propagated through space in the form of waves.

5.2. The Homogeneous Wave Equations

We have shown that the electric potential ϕ and the magnetic vector potential $\mathbf{A}$ representing a time-dependent electromagnetic field satisfy inhomogeneous wave equations. We obtained particular integrals of these equations, the retarded potentials, by the method due to Kirchhoff. However, to these particular solutions we can add any solution of the homogeneous wave equation as the complementary function. We now wish to consider in more detail these solutions of the homogeneous wave equation and their interpretation.

We can either consider the potentials ϕ and $\mathbf{A}$ which satisfy the equations

$$\nabla^2\phi - \frac{1}{c^2}\frac{\partial^2\phi}{\partial t^2} = 0 \tag{5.3}$$

and

$$\nabla^2\mathbf{A} - \frac{1}{c^2}\frac{\partial^2\mathbf{A}}{\partial t^2} = 0, \tag{5.4}$$

or we can work directly with the observable field intensities $\mathbf{E}$ and $\mathbf{B}$. We shall do the latter and first show that these intensities also satisfy homogeneous wave equations.

If we take the curl of the field equation

$$\nabla \times \mathbf{E} = -\frac{\partial \mathbf{B}}{\partial t}$$

we obtain, on interchanging the order of the operators,

$$\nabla \times (\nabla \times \mathbf{E}) = -\frac{\partial}{\partial t}(\nabla \times \mathbf{B}). \tag{5.5}$$

If we now use the relation

$$\nabla \times (\nabla \times \mathbf{E}) = \nabla(\nabla . \mathbf{E}) - \nabla^2\mathbf{E}$$

and restrict ourselves to regions free from charge and currents, we obtain, since for such regions $\nabla . \mathbf{E} = 0$,

$$\nabla^2\mathbf{E} = \frac{\partial}{\partial t}(\nabla \times \mathbf{B}). \tag{5.6}$$

Now let us take the time derivative of the field equation

$$\mathbf{\nabla} \times \mathbf{B} = \mu_0 \epsilon_0 \frac{\partial \mathbf{E}}{\partial t}$$

for this region to obtain

$$\frac{\partial}{\partial t} (\mathbf{\nabla} \times \mathbf{B}) = \frac{1}{c^2} \frac{\partial^2 \mathbf{E}}{\partial t^2}, \tag{5.7}$$

where as usual we have put $c = (\mu_0 \epsilon_0)^{-1/2}$.

Substituting from equation (5.7) into the right-hand side of equation (5.6) we obtain

$$\mathbf{\nabla}^2 \mathbf{E} = \frac{1}{c^2} \frac{\partial^2 \mathbf{E}}{\partial t^2}. \tag{5.8}$$

It can similarly be shown that

$$\mathbf{\nabla}^2 \mathbf{B} = \frac{1}{c^2} \frac{\partial^2 \mathbf{B}}{\partial t^2}. \tag{5.9}$$

Thus we have shown that in a region of space free from charge and currents, the electric and magnetic field intensities $\mathbf{E}$ and $\mathbf{B}$ satisfy homogeneous wave equations. This does *not* necessarily mean that $\mathbf{E}$ and $\mathbf{B}$ are propagated as waves, since any static solution of $\mathbf{\nabla}^2 \mathbf{E} = 0$ would satisfy equation (5.8) and we would not consider this to be a wave. What we must do is to actually construct a solution of equations (5.8) and (5.9) which has wavelike properties and which is also a solution of the field equations. While it is true that any $\mathbf{E}$ and $\mathbf{B}$ which satisfy the field equations also satisfy equations (5.8) and (5.9), the converse is not true, and whether a solution of equations (5.8) and (5.9) satisfies the field equations must be ascertained by direct substitution in the field equations.

5.3. Plane Electromagnetic Waves

In our considerations of electromagnetic fields propagating in space, we shall confine our attention to the simplest form of disturbance from the mathematical point of view, namely a plane sinusoidal wave.

Consider the electric and magnetic intensities defined by the equations

$$\left. \begin{aligned} \mathbf{E}(\mathbf{r}) &= \mathbf{E}_0 \sin (pt - \mathbf{k} . \mathbf{r}) \\ \mathbf{B}(\mathbf{r}) &= \mathbf{B}_0 \sin (pt - \mathbf{k} . \mathbf{r}) \end{aligned} \right\}, \tag{5.10}$$

where $\mathbf{E}_0$, $\mathbf{B}_0$ and $\mathbf{k}$ are constant vectors, and p is a constant scalar. We shall show that these field intensities can represent a plane electromagnetic wave travelling in the direction of the vector $\mathbf{k}$ and we shall determine the restrictions necessary for this to be so. The field intensities defined in equation (5.10) will satisfy the wave equations (5.8) and (5.9) only if

$$p/k = c. \tag{5.11}$$

This is our first restriction on the intensities. Applying the relation

$$\nabla \times (\phi \mathbf{A}) = \phi \nabla \times \mathbf{A} - \mathbf{A} \times \nabla \phi$$

to equations (5.10) we see that

$$\left.\begin{array}{l} \nabla \times \mathbf{E} = -\mathbf{k} \times \mathbf{E}_0 \cos{(pt - \mathbf{k} \cdot \mathbf{r})} \\ \nabla \times \mathbf{B} = -\mathbf{k} \times \mathbf{B}_0 \cos{(pt - \mathbf{k} \cdot \mathbf{r})} \end{array}\right\}. \tag{5.12}$$

If we differentiate equations (5.10) with respect to time we obtain

$$\left.\begin{array}{l} \dfrac{\partial \mathbf{E}}{\partial t} = p\mathbf{E}_0 \cos{(pt - \mathbf{k} \cdot \mathbf{r})} \\[2mm] \dfrac{\partial \mathbf{B}}{\partial t} = p\mathbf{B}_0 \cos{(pt - \mathbf{k} \cdot \mathbf{r})} \end{array}\right\}. \tag{5.13}$$

Hence if we are to satisfy the field equations

$$\left.\begin{array}{l} \nabla \times \mathbf{E} = -\dfrac{\partial \mathbf{B}}{\partial t} \\[2mm] \nabla \times \mathbf{B} = \dfrac{1}{c^2} \dfrac{\partial \mathbf{E}}{\partial t} \end{array}\right\}, \tag{5.14}$$

we must have the restrictions

$$\left.\begin{array}{l} \mathbf{k} \times \mathbf{E}_0 = p\mathbf{B}_0 \\ c^2 \mathbf{k} \times \mathbf{B}_0 = -p\mathbf{E}_0 \end{array}\right\}. \tag{5.15}$$

These imply that the vectors $\mathbf{k}$, $\mathbf{E}_0$ and $\mathbf{B}_0$ form an orthogonal triad in the right-handed sense. We also see from equations (5.15) and (5.11) that in our units $|\mathbf{E}_0| = c|\mathbf{B}_0|$.

If we look at the intensities (5.10) we see that they are constant over the surfaces represented by

$$pt - \mathbf{k} \cdot \mathbf{r} = \text{constant}, \tag{5.16}$$

which for a fixed time t reduces to

$$\mathbf{k.r} = \text{constant}; \tag{5.17}$$

they are the equations of planes normal to the direction of the vector **k**. The constant in equation (5.17) is proportional to the length of the perpendicular from the origin on to this plane (Fig. 8). Now

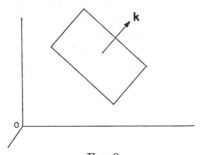

FIG. 8

returning to the general case of equation (5.16), we see that this represents a plane normal to **k** moving in the direction of **k**. The velocity with which the waves move is p/k which is equal to c.

In describing plane waves we use the following nomenclature. We call p the angular frequency of the waves, or $p/2\pi$ the frequency, since the wave pattern repeats itself $p/2\pi$ times per second at any fixed point. We call the vector **k** the wave vector, and its modulus k the wave number. The wave vector **k** gives the direction of the normal to the wave-fronts or planes over which the intensities are constant. The distance $2\pi/k$ is called the wavelength since the wave pattern repeats itself after this distance at any fixed time. The velocity p/k or c is referred to as the phase velocity of the wave. The main reason for introducing the constant c into the field equations was so that the equations we obtained for the field intensities would take the well known form of wave equations.

5.4. THE ELECTROMAGNETIC THEORY OF LIGHT

In the last section we have shown that a possible solution of the field equations corresponds to a plane wave moving *in vacuo* with a phase velocity given by the constant c which appears in the field equations. This velocity can be obtained from the results of experiments in electromagnetism and has the value of approximately

3×10^8 m/sec. When Maxwell deduced this possible solution he was struck by the agreement of this velocity c and the known velocity of light *in vacuo*. He postulated on these theoretical grounds that light was a form of electromagnetic radiation. This conjecture was experimentally verified several years later by Hertz. We now know that there are several forms of electromagnetic radiation, which we distinguished as X-rays, ultra-violet, visible, infra-red or radio waves according to their frequency. We refer to the whole range of possible waves, from the short-wavelength X-rays to the long-wavelength radio waves as the *electromagnetic spectrum*.

This discovery by Maxwell brought the whole field of optics within the realm of electromagnetic theory. This was a great unification of subjects and placed the science of optics on a much more secure foundation than it had hitherto possessed.

If one takes into account the electromagnetic properties of matter, one can also obtain the laws of optics for media from the field equations; however we shall not consider this here.

Now let us consider a few simple examples involving plane electromagnetic waves in order to give us some facility in handling the quantities involved.

▶ *Example*

The magnetic intensity in a plane electromagnetic wave is

$$\{a \cos [n(t - (y \sin \theta + z \cos \theta)/c)], \, 0, \, 0\}.$$

Find the electric intensity.

▶ *Solution*

Since the homogeneous field equations are linear, we can write both the electric and magnetic intensities as complex quantities if we wish; then we can take real parts of these complex solutions, or imaginary parts if we wish, to obtain the actual fields. This is a technique often used, since writing the intensities as complex quantities often makes the solving of the equations very much easier.

We write

$$\mathbf{B} = (a, \, 0, \, 0) \exp [in \, (t - \mathbf{k}.\mathbf{r})], \tag{1}$$

$$\mathbf{E} = \mathbf{E}_0 \exp [in \, (t - \mathbf{k}.\mathbf{r})], \tag{2}$$

with

$$\mathbf{k} = c^{-1}(0, \, \sin \theta, \, \cos \theta), \qquad k = c^{-1}. \tag{3}$$

Then in this case we wish to calculate the real part of $\mathbf{E}$, since the actual magnetic field corresponds to the real part of $\mathbf{B}$.

From the field equation

$$\nabla \times \mathbf{E} = -\frac{\partial \mathbf{B}}{\partial t},$$

we obtain

$$\mathbf{B} = \mathbf{k} \times \mathbf{E}.$$

Thus

$$\mathbf{k} \times (\mathbf{k} \times \mathbf{E}) = \mathbf{k} \times \mathbf{B}, \tag{4}$$

and since we assume no charges to be present,

$$\nabla . \mathbf{E} = 0.$$

This implies that

$$\mathbf{k} . \mathbf{E} = 0.$$

Therefore equation (4) becomes

$$k^2 \mathbf{E} = \mathbf{B} \times \mathbf{k}$$

and hence

$$\mathbf{E} = (\mathbf{B} \times \mathbf{k})/k^2.$$

Now to obtain our required field we must take the real part of this expression:

$$\mathbf{E} = c(0, \, -a \cos \theta, \, a \sin \theta) \cos \{n[t - (y \sin \theta + z \cos \theta)/c]\}.$$

To this solution we can add any static electric field and we still have a solution of the field equations, but this extra static field will not be connected with a wave motion. ◄

Now let us consider an example which will illustrate how a plane electromagnetic wave behaves in a conducting medium, where the relationship between electric intensity and current density is often linear.

▶ *Example*

Show that if a plane electromagnetic wave is propagating in a region where we can write $\mathbf{j} = \sigma \mathbf{E}$, then as the wave progresses the amplitude decreases if σ is a positive constant.

▶ *Solution*

In the present situation when we have $j = \sigma E$ we can assume that the density of free charge is zero. This was shown to be so in the first problem after Chapter 2.

If we take the curl of the field equation

$$\nabla \times E = -\frac{\partial B}{\partial t}, \tag{1}$$

we obtain

$$\nabla \times (\nabla \times E) = \nabla(\nabla . E) - \nabla^2 E = -\nabla \times \frac{\partial B}{\partial t} = -\frac{\partial}{\partial t}(\nabla \times B). \tag{2}$$

If we now substitute for $\nabla \times B$ in equation (2) from the field equation

$$\nabla \times B = \mu_0 \left(j + \epsilon_0 \frac{\partial E}{\partial t} \right), \tag{3}$$

and use the fact that

$$\nabla . E = 0 \quad \text{and} \quad j = \sigma E,$$

we obtain

$$\nabla^2 E = \mu_0 \sigma \frac{\partial E}{\partial t} + \frac{1}{c^2} \frac{\partial^2 E}{\partial t^2} \tag{4}$$

which is a wave equation.

We now consider a solution of equation (4) which could represent a plane wave travelling in the $-x$ direction. We put, for $x < 0$,

$$E = E_0 \exp[i(nt + kx)] \tag{5}$$

which on substitution into equation (4) gives us a relation between the angular frequency n of the wave and the wave number k. This is

$$-k^2 = \mu_0 \sigma i n - n^2/c^2 \tag{6}$$

or

$$k = \frac{n}{c} \left(1 - \frac{i\sigma}{n\epsilon_0} \right)^{1/2} \tag{7}$$

which we can write in the form

$$k = \alpha - i\beta$$

with $\beta > 0$.

We choose this root of equation (6) so that in the limiting case of zero conductivity, that is when $\sigma = 0$ as we have *in vacuo*, the root becomes n/c representing a plane wave *in vacuo* which advances in the negative x-direction. Then the solution of the field equations for such a plane wave is

$$\left.\begin{array}{l} \mathbf{E} = \mathbf{E}_0 \exp\left[i(nt + \alpha x)\right] \exp\left(\beta x\right) \\ \mathbf{B} = (\mathbf{k}' \times \mathbf{E})(\alpha - i\beta)/nk' \end{array}\right\} \quad x < 0,$$

where the vector $\mathbf{k}'$ is in the positive x-direction.

From these results we see that the β term causes the wave amplitude to decrease as the wave progresses towards $x = -\infty$. We also see that the wave is made up to two intensities $\mathbf{E}$ and $\mathbf{B}$ which are no longer in phase; they reach their extreme magnitudes at different places for any fixed time.

As stated earlier, the phenomena which are illustrated in this example give us a good idea of what happens when a plane wave travels in a metal. ◄

PROBLEMS FOR CHAPTER 5

(1) The electric intensity in a plane wave is

$$\mathbf{E} = \mathbf{E}_0 \sin\left(pt - \mathbf{k}.\mathbf{r}\right),$$

where $\mathbf{E}_0$ is a constant vector. Find the magnetic intensity, assuming no charges or currents present.

(2) The electric intensity in a plane wave is

$$\mathbf{E} = [0, A \cos\left(pt - kx\right), A \sin\left(pt - kx\right)],$$

where A is a constant. Find the magnetic intensity. [Such a wave in which the end-point of the electric vector describes a circle is said to be circularly polarised. In the previous example the locus of the end-point of the electric intensity described a straight line and we say the wave is linearly polarised.]

(3) Show that if we choose a solution of the homogeneous wave equation of the form

$$\phi(x, y, z, t) = \phi_0 \exp\left\{i[\omega t + kS(x, y, z)]\right\}$$

with $k = \omega/c$ and ϕ_0 a constant, then the phase function S satisfies the equation

$$(\nabla S)^2 - (i/k)\nabla^2 S = 1.$$

[This equation is used in the discussion of the relationship between geometrical and wave optics.]

Chapter 6

ENERGY AND MOMENTUM OF AN ELECTROMAGNETIC FIELD

6.1. Reasons for Energy and Momentum Considerations in an Electromagnetic Field

Let us now turn our attention to the problems of how energy and momentum are propagated and stored in an electromagnetic field. We can easily see that energy is propagated if we consider the following situation. Suppose a current starts flowing in a wire. To do this we have to provide an electric field to cause the electrons in the wire to move. The production of this electric field requires work from the generator. The wire then radiates an electromagnetic field which can produce a force on a charged particle causing it to accelerate and increase its kinetic energy. Thus we have a transfer of energy from the generator producing the current in the wire to the charged particle.

It has been established experimentally that when an electromagnetic wave hits a body it exerts a pressure on it which will cause a change in the momentum of the body. Hence, if we are to have momentum conserved we must somehow associate momentum with the electromagnetic wave so that the gain in momentum of the body can be accounted for by the loss in momentum of the wave when it hits the body.

6.2. The Poynting Vector

If we take the scalar product of the field equation

$$\nabla \times \mathbf{B} - \frac{1}{c^2}\frac{\partial \mathbf{E}}{\partial t} = \mu_0 \mathbf{j}$$

with $\mathbf{E}$, and subtract from it the scalar product of the field equation

$$\nabla \times \mathbf{E} + \frac{\partial \mathbf{B}}{\partial t} = 0$$

with **B**, we obtain

$$\mathbf{E}.\left(\boldsymbol{\nabla}\times\mathbf{B}-\frac{1}{c^2}\frac{\partial\mathbf{E}}{\partial t}\right)-\mathbf{B}.\left(\boldsymbol{\nabla}\times\mathbf{E}+\frac{\partial\mathbf{B}}{\partial t}\right) = \mu_0\mathbf{E}.\mathbf{j}. \qquad (6.1)$$

Using the identity

$$\mathbf{B}.(\boldsymbol{\nabla}\times\mathbf{E})-\mathbf{E}.(\boldsymbol{\nabla}\times\mathbf{B}) = \boldsymbol{\nabla}.(\mathbf{E}\times\mathbf{B}), \qquad (6.2)$$

we can write equation (6.1) in the form

$$\boldsymbol{\nabla}.(\mathbf{E}\times\mathbf{B})+\mu_0\mathbf{E}.\mathbf{j} = -\frac{1}{c^2}\,\mathbf{E}\cdot\frac{\partial\mathbf{E}}{\partial t}-\mathbf{B}\cdot\frac{\partial\mathbf{B}}{\partial t}. \qquad (6.3)$$

Suppose that we have a region of space V, bounded by a surface S, in which we are trying to describe the processes of energy and momentum transfer by the electromagnetic field. Let us integrate both sides of equation (6.3) over this region V, obtaining

$$\iiint_V \boldsymbol{\nabla}.(\mathbf{E}\times\mathbf{B})\,d\tau+\iiint_V \mu_0\mathbf{E}.\mathbf{j}\,d\tau = -\frac{\partial}{\partial t}\iiint_V \frac{1}{2}\left(\frac{E^2}{c^2}+B^2\right)d\tau, \quad (6.4)$$

where we have brought the differential operator on the left-hand side from under the integral sign and rearranged some of the terms. If we now apply the divergence theorem to the first term of equation (6.4), we obtain the very important result known as *Poynting's Theorem*[*]:

$$\frac{1}{\mu_0}\iint_S (\mathbf{E}\times\mathbf{B}).d\mathbf{S}+\iiint_V \mathbf{E}.\mathbf{j}\,d\tau = -\frac{\partial}{\partial t}\iiint_V \frac{1}{2\mu_0}\left(\frac{E^2}{c^2}+B^2\right)d\tau. \quad (6.5)$$

We give the various terms which appear in this equation the following physical interpretation. First consider the second term on the left-hand side. We take this to represent the rate at which work is done in driving the current in the region, for the following reason. If we have a charge q moving in an electric field **E** the work done by the field in moving this charge through an infinitesimal displacement **dr** is $q\mathbf{E}.\mathbf{dr}$. Hence if we have N such charges, the work done is $Nq\mathbf{E}.\mathbf{dr}$. If we have N charges per unit volume moving with velocity **v**, then the work done per unit volume in the infinitesimal time dt is $Nq\mathbf{E}.\mathbf{v}\,dt$, which gives a rate of working of the field per unit volume of $Nq\mathbf{E}.\mathbf{v}$. But $Nq\mathbf{v}=\mathbf{j}$, and we have the required result that

---

[*] J. H. Poynting (1852–1914), the British physicist who first gave explicit general formulae for energy and momentum calculations in electromagnetic fields.

the rate of working per unit volume is $\mathbf{E} \cdot \mathbf{j}$, giving the stated interpretation of the second term on the left-hand side of equation (6.5).

We interpret the right-hand side of equation (6.5) as the rate of
increase in the energy content of the electromagnetic field in the
volume V. This suggests that we regard the quantity $B^2/2\mu_0$ as the
energy density of the magnetic field and the quantity $\epsilon_0 E^2/2$ as
the energy density of the electric field. These assumptions are well
supported by experimental evidence.

The left-hand side of equation (6.5) is a surface integral and this
suggests, in conjunction with the aim of obtaining an equation which
represents an energy balance, that we treat it as the rate of flow of
energy out of the volume V through the surface S. We call the vector

$$\mathbf{P} = (\mathbf{E} \times \mathbf{B})/\mu_0 \qquad (6.6)$$

the *Poynting vector*. We must always use the energy balance equation
(6.5) in the integral form as we have derived it. It is not generally
possible to regard the Poynting vector as an energy flow density
(watts/m^2) out of V across S. As an illustration we could have static
electric and magnetic fields at right angles. This arrangement gives
a nonzero Poynting vector, but there is no energy flow out of any
closed surface. Of course if we use the integral form (6.5) we obtain
zero energy flow out of any such closed surface S.

If we give to each term in equation (6.5) the physical interpretation
we have discussed, then we can regard this equation as illustrating
an energy balance or conservation theorem. It is a most useful result
and gives a deeper insight into the nature of electromagnetic
phenomena.

6.3. THE ELECTROMAGNETIC STRESS TENSOR

We now consider another aspect of energy and momentum in an
electromagnetic field, and discuss how the change in these quantities
for a given region can be described in terms of 'stresses' which act on
the surface which bounds the region. These stresses were first
formulated by Maxwell, and as well as being a useful tool in certain
circumstances they are used in the development of the four-dimensional relativistic formulation of electrodynamics.

Suppose that we have a region V bounded by a closed surface S.
Suppose that in V there is a charge distribution with density $\rho(\mathbf{r})$.
What we wish to consider is the rate of change of the total momentum

6

of the region, mechanical and electromagnetic, under electromagnetic forces. As we have said we shall try to represent this change of momentum as due to some system of forces acting across S, in the same way one discusses stresses and strains in an elastic solid.

Let $\mathbf{G}$ represent the mechanical momentum of the matter in the region V. The rate of change of this momentum is the total force on V due to any electromagnetic fields in V, i.e.

$$\frac{d\mathbf{G}}{dt} = \int\!\!\!\int\!\!\!\int_V \rho[\mathbf{E} + (\mathbf{v} \times \mathbf{B})]\, d\tau. \tag{6.7}$$

Here square brackets do not denote retarded values. If we write the current $\mathbf{j}$ in terms of the motion of the charge, then on the microscopic scale we have $\mathbf{j} = \rho\mathbf{v}$, and the field equations take the form given by Lorentz

$$\mathbf{\nabla}.\mathbf{E} = \rho/\epsilon_0, \tag{6.8}$$

$$\mathbf{\nabla} \times \mathbf{B} - \frac{1}{c^2}\frac{\partial \mathbf{E}}{\partial t} = \mu_0 \rho\mathbf{v}. \tag{6.9}$$

Inserting these field equations into equation (6.7), we obtain

$$\frac{d\mathbf{G}}{dt} = \int\!\!\!\int\!\!\!\int_V \left[\epsilon_0(\mathbf{\nabla}.\mathbf{E})\mathbf{E} + \frac{[(\mathbf{\nabla} \times \mathbf{B}) \times \mathbf{B}]}{\mu_0} - \epsilon_0\left(\frac{\partial \mathbf{E}}{\partial t} \times \mathbf{B}\right)\right] d\tau,$$

which we then rewrite in the form

$$\frac{d\mathbf{G}}{dt} = \int\!\!\!\int\!\!\!\int_V \left[\epsilon_0(\mathbf{\nabla}.\mathbf{E})\mathbf{E} + \frac{(\mathbf{\nabla} \times \mathbf{B}) \times \mathbf{B}}{\mu_0} \right.$$
$$\left. + \epsilon_0\mathbf{E} \times \frac{\partial \mathbf{B}}{\partial t} - \epsilon_0\frac{\partial}{\partial t}(\mathbf{E} \times \mathbf{B})\right] d\tau. \tag{6.10}$$

If we now use the field equation

$$\mathbf{\nabla} \times \mathbf{E} = -\frac{\partial \mathbf{B}}{\partial t},$$

we can write equation (6.10) in the form

$$\frac{d\mathbf{G}}{dt} + \frac{\partial}{\partial t}\int\!\!\!\int\!\!\!\int_V \mathbf{g}\, d\tau = \int\!\!\!\int\!\!\!\int_V \left[\epsilon_0(\mathbf{\nabla}.\mathbf{E})\mathbf{E} + \frac{(\mathbf{\nabla} \times \mathbf{B}) \times \mathbf{B}}{\mu_0} + \epsilon_0(\mathbf{\nabla} \times \mathbf{E}) \times \mathbf{E}\right] d\tau,$$

$$\tag{6.11}$$

where

$$\mathbf{g} = \epsilon_0(\mathbf{E} \times \mathbf{B}).$$

If we now define the array of quantities p_{ij} by the relations

$$\left.\begin{array}{l} p_{ii} = -\tfrac{1}{2}[\epsilon_0(E_i^2 - E_j^2 - E_k^2) + (B_i^2 - B_j^2 - B_k^2)/\mu_0] \\ p_{ij} = -\tfrac{1}{2}(\epsilon_0 E_i E_j + B_i B_j/\mu_0) \end{array}\right\}, \quad (6.12)$$

where we have used no summation and the indices i, j and k run from 1 to 3 and are all different, then we can write equation (6.11) in the form

$$\frac{dG_i}{dt} + \frac{\partial}{\partial t}\int\!\!\int\!\!\int_V g_i \, d\tau = -\int\!\!\int\!\!\int_V \frac{\partial p_{ij}}{\partial x^j} \, d\tau. \quad (6.13)$$

If we now use the divergence theorem to simplify the right-hand side, this equation assumes a form which is more easily interpreted, i.e. we put

$$-\int\!\!\int\!\!\int_V \left(\frac{\partial p_{ix}}{\partial x} + \frac{\partial p_{iy}}{\partial y} + \frac{\partial p_{iz}}{\partial z}\right) dx \, dy \, dz = -\int\!\!\int_S [p_{ix}]_{x_1}^{x_2} + \cdots \quad (6.14)$$

where x_1 and x_2 are the x-co-ordinates of the point of intersection of a line parallel to the x-axis with S.

We interpret the left-hand side of equation (6.13) as follows. The first term clearly represents the rate of change of the ith component of the mechanical momentum of the matter within V. The second term is interpreted as the rate of change of the corresponding component of the momentum of the electromagnetic field in the region V. If we thus regard the left-hand side of equation (6.13) as the rate of change of the total momentum of the system, we see that it can be related to the surface integrals of the quantities p_{ij} over the surface S. We therefore interpret this array as producing a force on the system within V and think of the action as being that of a stress acting across the surface S. Again we must really restrict ourselves to using the integral form of the equation.

The force $\mathbf{F}$ which acts across a closed surface S is given by

$$F_i = \int\!\!\int_S p_{ij} \, dS_j. \quad (6.15)$$

We have shown that we can regard an electromagnetic field as producing stresses in the region it occupies, and that we can use these stresses to calculate the force acting on any volume. For instance, if we consider the case of a region V containing a uniform electric field we can obtain the effect on the region outside V by considering the region to be made of material which exerts a tension $\epsilon_0 E^2/2$ parallel to the direction of the field and a pressure $\epsilon_0 E^2/2$ normal to the direction of the field.

Now let us consider an example to illustrate the results of this section.

▶ *Example*

Show that if we have a closed surface containing radiation which is homogeneous and isotropic, then the pressure exerted on the surface S is equal to one-third of the energy density of the radiation.

▶ *Solution*

Consider an element dS of the surface S and take the x-axis normal to the element. Since the radiation is homogeneous and isotropic, the force acting on the element will be along the normal. Using equation (6.15) we see that this force is $p_{xx} \, dS$.

From the isotropy and homogeneity of the radiation we can put:

$$E_x^2 = E_y^2 = E_z^2 = E^2/3,$$
$$B_x^2 = B_y^2 = B_z^2 = B^2/3.$$

If we now use these equations with equation (6.12) we see that at dS

$$p_{xx} = (\epsilon_0 E^2 + B^2/\mu_0)/6 = U/3,$$

where U is the energy density of the radiation. This is the required result.   ◀

The radiation inside a surface will only be completely isotropic and homogeneous if the surface S absorbs all the incident radiation. Hence the supply of radiation must be renewed.

One can by a similar method treat the case when the surface S is a perfect reflector of the radiation, but we shall not discuss this case here. Let us, however, take another example of some interest.

▶ *Example*

A plane electromagnetic wave has field intensities

$$\mathbf{E} = A \cos{(pt - kz)}\mathbf{i},$$
$$\mathbf{B} = (A/c) \cos{(pt - kz)}\mathbf{j}.$$

where $p/k = c$, and $\mathbf{i}$ and $\mathbf{j}$ are unit vectors along the x- and y-axes respectively. Determine the distribution of energy in the wave.

▶ *Solution*

The energy density $U(z)$ in the wave is

$$U(z) = \frac{1}{2\mu_0}\left(\frac{E^2}{c^2} + B^2\right) = \frac{A^2 \cos^2(pt - kz)}{\mu_0 c^2}$$

$$= \epsilon_0 A^2 \cos^2(pt - kz).$$

The amount of energy contained in a prism of unit area cross-section and one wavelength, say λ, long with generators parallel to the x-axis is

$$Q = \int\limits_{z}^{z+\lambda} U(z)\,dz = \int\limits_{z}^{z+(2\pi/k)} U(z)\,dz = \epsilon_0 \pi A^2/k. \qquad \blacktriangleleft$$

▶ *Example*

Suppose the rate at which energy is absorbed by a square metre of the earth's surface from sunlight is 1500 W. Determine the amplitude of the electric intensity of the electromagnetic waves assuming them to be sinusoidal and plane.

▶ *Solution*

Let us assume that the normal to the surface is in the z direction. The waves will fall so that their wave vectors $\mathbf{k}$ are along the normal. Let the electric intensity in a plane wave be

$$\mathbf{E} = \mathbf{E}_0 \sin(pt - \mathbf{k}\cdot\mathbf{r})$$

where $\mathbf{E}_0$ is a constant vector, perpendicular to $\mathbf{k}$. From the results of Problem (1) of Chapter 5 we can write for the magnetic intensity

$$\mathbf{B} = (\mathbf{k}\times\mathbf{E}_0)\sin(pt - \mathbf{k}\cdot\mathbf{r})/kc.$$

Hence the Poynting vector is

$$\mathbf{P} = (\mathbf{E}\times\mathbf{B})/\mu_0$$

$$= \mathbf{E}_0 \times(\mathbf{k}\times\mathbf{E}_0)\sin^2(pt - \mathbf{k}\cdot\mathbf{r})/ck\mu_0$$

$$= E_0^2 \sin^2(pt - \mathbf{k}\cdot\mathbf{r})\mathbf{k}/ck\mu_0$$

$$= \epsilon_0 c E_0^2 \sin^2(pt - \mathbf{k}\cdot\mathbf{r})\mathbf{k}/k.$$

If we consider the radiation at a fixed place over a period of a second, the phase factor $(pt - \mathbf{k.r})$ passes through many multiples of π and the factor $\sin^2 (pt - \mathbf{k.r})$ gives a mean value of $\frac{1}{2}$. Thus if we equate the time average of the Poynting vector to the rate of energy absorption we obtain

$$1500 = (8\cdot854 \times 10^{-12})(3 \times 10^8)(0.5)E_0^2,$$

and hence the amplitude of the electric intensity, E_0, is given by

$$E_0 = 1\cdot063 \times 10^3 \text{ V/m}.$$

We have used the result that the Poynting vector for each plane wave, and hence for the total radiation, is parallel to the normal of the surface.   ◀

▶ *Example*

Using the data of the previous example determine the radiation pressure due to the sun at the surface of the earth.

▶ *Solution*

Let us take a surface with its normal in the direction of the z-axis and consider waves falling such that their wave vectors are in the z-direction. The force on this surface due to the radiation can be written, using equation (6.15), as

$$F_i = p_{ij}n_j,$$

where we have assumed that the radiation stress tensor p_{ij} is constant over the surface which has direction cosines n_j. Thus the normal pressure on the surface is

$$F_z = p_{zz}n_z.$$

Hence

$$F_z = \frac{1}{2}(\epsilon_0 E^2 + B^2/\mu_0)$$
$$= \frac{1}{2}(\epsilon_0 E_0^2 + E_0^2/c^2\mu_0) \sin^2 (pt - \mathbf{k.r})$$
$$= \epsilon_0 E_0^2 \sin^2 (pt - \mathbf{k.r}).$$

Again the last term produces an average value of $\frac{1}{2}$ and hence the pressure is given by

$$F_z = (8\cdot854 \times 10^{-12})(1\cdot063)^2 \times 10^6 \times (0.5)$$
$$= 4\cdot96 \times 10^{-6} \text{ newtons/m}^2.$$ ◀

Problems for Chapter 6

(1) A steady current J flows in a uniform straight wire due to an electric field $\mathbf{E}$ parallel to the wire. Show that the rate at which energy flows per unit length into the wire from the field is the same as the rate at which work is done on the current in this length of wire.

(2) A magnetic dipole of moment $\mathbf{m}$ lies along the x-axis at the origin. An electric dipole of moment $\mathbf{M}$ lies along the y-axis at the origin. Show that the flux of the Poynting vector out of any closed surface which does not meet the origin is zero.

(3) Suppose we have two concentric spherical conductors carrying equal and opposite charges. Show that their mutual potential energy, as defined in Chapter 3, is equal to the energy residing in the electric field between them.

Chapter 7

MOTION OF CHARGED PARTICLES IN ELECTROMAGNETIC FIELDS

7.1. The Newtonian Equations of Motion of a Charged Particle

In this chapter we shall consider the motion of a charged particle in various types of electromagnetic field. We shall assume that the force which acts on the particle is the Lorentz force

$$\mathbf{F} = q[\mathbf{E} + (\mathbf{v} \times \mathbf{B})] \tag{7.1}$$

in our usual notation. We shall also assume that the fields $\mathbf{E}$ and $\mathbf{B}$ are given, and thus we ignore for the moment the very difficult problem of how the fields produced by the particle itself affect its motion. We shall return to this much more difficult problem in Chapter 9.

We shall confine our attention to nonrelativistic particles, that is to particles whose velocities are very much less than the velocity of light. This avoids the complexities of equations of motion in which the mass of a particle is a function of its velocity. We can under these conditions write the equations of motion for a charge q, with mass m, as

$$m \frac{d\mathbf{v}}{dt} = q[\mathbf{E} + (\mathbf{v} \times \mathbf{B})]. \tag{7.2}$$

We shall now consider several simple field configurations and discuss the motions in each.

(a) Uniform Electric Field

For a uniform electric field the equation of motion (7.2) reduces to

$$m \frac{d\mathbf{v}}{dt} = q\mathbf{E} \tag{7.3}$$

which gives on integration, if $\mathbf{E}$ is uniform and time-independent,

$$\mathbf{v} = (q\mathbf{E}/m)t + \mathbf{V},$$

where **V** is a constant vector to be determined from the initial conditions of the problem. A second integration gives

$$\mathbf{r} = (qt^2/2m)\mathbf{E} + t\mathbf{V} + \mathbf{W}, \qquad (7.4)$$

where again **W** is a constant vector determined from the initial conditions.

According to our equation (7.3), a charged particle in a uniform constant electric field should accelerate indefinitely at a finite rate. This implies that ultimately the particle can have an indefinitely large velocity. If one treats the same problem relativistically however, we find that the velocity never exceeds the velocity of light. The increase of mass with velocity prevents this from happening. In practice of course we cannot produce uniform electric fields of any great extent.

(b) Magnetic Field

Suppose that we have a particle moving in a magnetic field only. In this case the equation of motion becomes

$$m \frac{d\mathbf{v}}{dt} = q\mathbf{v} \times \mathbf{B}. \qquad (7.5)$$

If we take the scalar product of both sides of this equation with **v** we obtain

$$m\mathbf{v} \cdot \frac{d\mathbf{v}}{dt} = q(\mathbf{v} \times \mathbf{B}) . \mathbf{v} = 0, \qquad (7.6)$$

since the scalar triple product has a repeated factor. We can rewrite this equation in terms of the kinetic energy T of the charged particle as

$$m\mathbf{v} \cdot \frac{d\mathbf{v}}{dt} = \frac{d}{dt} \left(\tfrac{1}{2}mv^2 \right) = \frac{dT}{dt} = 0.$$

Hence we see that in a magnetic field the energy of a charged particle is a constant of the motion. We cannot use a static magnetic field to change the speed of charged particles, but only their direction of motion. With a time-dependent magnetic field there is an associated electric field, which will change the energy of the particle.

(c) Static Uniform Magnetic Field

If we have a static uniform magnetic field there is no associated electric field, and the equation of motion is simply

$$m \frac{d\mathbf{v}}{dt} = q\mathbf{v} \times \mathbf{B} \qquad (7.7)$$

with **B** constant in space and time and v constant in time.

From the equation of motion (7.7) we see that there is no force component parallel to the direction of the magnetic intensity **B**, and hence any initial motion in this direction continues unchanged. In view of this fact, we restrict the motion to a plane perpendicular to the direction of **B**. The magnitude of the acceleration is constant, since v and B are constant and **v** is perpendicular to **B**, and the direction of the acceleration is normal to **B** and **v**. This implies that the particle moves in a circle. Thus the resultant motion is the superposition of this circular motion and the translation parallel to **B** from the initial velocity, i.e. the particle moves in a circular helix of constant pitch with axis in the direction of the magnetic intensity.

The angular velocity of the particle about the axis of the helix is the so-called 'gyrofrequency' of the particle, say Ω, where

$$\Omega = qB/m. \tag{7.8}$$

The radius of the helix is

$$R = mv/qB, \tag{7.9}$$

which is a constant.

(d) Perpendicular Uniform Electric and Magnetic Fields

In this more complicated configuration we have

$$m \frac{d\mathbf{v}}{dt} = q[\mathbf{E} + (\mathbf{v} \times \mathbf{B})], \tag{7.10}$$

with the restriction of static uniform **E** and **B**, and

$$\mathbf{E} \cdot \mathbf{B} = 0. \tag{7.11}$$

Let us simplify the problem by the following transformation. Put

$$\mathbf{v}' = \mathbf{v} - (\mathbf{E} \times \mathbf{B})/B^2. \tag{7.12}$$

In terms of the new velocity **v**′ the equation of motion becomes

$$m \frac{d\mathbf{v}'}{dt} = q \left[\mathbf{E} + (\mathbf{v}' \times \mathbf{B}) + \frac{[(\mathbf{E} \times \mathbf{B}) \times \mathbf{B}]}{B^2} \right].$$

For the fields we are considering we can write

$$(\mathbf{E} \times \mathbf{B}) \times \mathbf{B} = (\mathbf{E} \cdot \mathbf{B})\mathbf{B} - B^2\mathbf{E} = -B^2\mathbf{E},$$

and equation (7.13) reduces to

$$m \frac{d\mathbf{v}'}{dt} = q\mathbf{v}' \times \mathbf{B}, \tag{7.13}$$

which is an equation whose solution we have discussed in Section (c). Thus in the present case we have the following motion. The motion parallel to **B** is one of constant velocity. The motion in a direction perpendicular to **B** is, from equations (7.12) and (7.13), motion in a circle about a centre which is moving normal to **E** and **B** with a constant velocity $(\mathbf{E} \times \mathbf{B})/B^2$. We call this moving centre the 'guiding centre' for the motion. Again the particle describes the circular motion with the gyrofrequency (from equation (7.9)).

If the field configuration becomes very much more complicated than those we have described in the last few sections, the problems become intractable by analytic methods. For instance, the very important case of the motion of charges in a magnetic dipole field is extremely difficult to treat completely. This problem is relevant to the motion of the charged particles in the cosmic radiation which reach the earth from space and interact with the earth's magnetic field.

We now consider the problem of the motion of charged particles in an electromagnetic field from an alternative, and often more useful, point of view.

7.2. The Lagrangian of a Charged Particle in an Electromagnetic Field

In the previous section we have discussed the motion of a charged particle in an electromagnetic field with equations of motion given by direct application of Newton's laws. Often it is more convenient, both in practical applications of the theory and in more basic discussions, to use the alternative approach of analytical mechanics and to try and apply the Lagrangian* equations of motion for a system.

Lagrange's equations of motion for a particle in a field would take the form

$$\frac{d}{dt}\left(\frac{\partial T}{\partial \dot{q}_r}\right) - \frac{\partial T}{\partial q_r} = Q_r, \tag{7.14}$$

where T is the kinetic energy of the system in terms of the generalised co-ordinates q_r of the system, and Q_r are the generalised forces acting on the particle.

---

* Called after J. L. Lagrange (1736–1873), one of the greatest pioneer workers in classical mechanics.

In the special case when the generalised forces Q_r can be written in the form

$$Q_r = \frac{d}{dt}\left(\frac{\partial M}{\partial \dot{q}_r}\right) - \frac{\partial M}{\partial q_r} \tag{7.15}$$

for some function $M(q_r, \dot{q}_r)$, then equation (7.14) takes the form

$$\frac{d}{dt}\left(\frac{\partial L}{\partial \dot{q}_r}\right) - \frac{\partial L}{\partial q_r} = 0, \tag{7.16}$$

where L is called the *Lagrangian* of the system and is defined by

$$L = T - M. \tag{7.17}$$

We shall show that for an electromagnetic field the generalised forces acting on a charged particle can indeed be put into the form (7.15). We shall now determine the appropriate M function and hence the Lagrangian.

In vacuo we can express the field intensities $\mathbf{E}$ and $\mathbf{B}$ as functions of the potentials ϕ and $\mathbf{A}$:

$$\left.\begin{array}{l} \mathbf{B} = \mathbf{\nabla} \times \mathbf{A} \\ \mathbf{E} = -\mathbf{\nabla}\phi - \dfrac{\partial \mathbf{A}}{\partial t} \end{array}\right\}. \tag{7.18}$$

The force on a charged particle is the Lorentz force

$$\mathbf{F} = q[\mathbf{E} + \mathbf{v} \times \mathbf{B}]$$

$$= q\left[-\mathbf{\nabla}\phi - \frac{\partial \mathbf{A}}{\partial t} + \mathbf{v} \times (\mathbf{\nabla} \times \mathbf{A})\right].$$

But

$$\mathbf{v} \times (\mathbf{\nabla} \times \mathbf{A}) = \mathbf{\nabla}(\mathbf{v}.\mathbf{A}) - (\mathbf{v}.\mathbf{\nabla})\mathbf{A},$$

and hence we can write the Lorentz force as

$$\mathbf{F} = \frac{d}{dt}(m\mathbf{v}) = q\left[-\mathbf{\nabla}\phi - \frac{\partial \mathbf{A}}{\partial t} - (\mathbf{v}.\mathbf{\nabla})\mathbf{A} + \mathbf{\nabla}(\mathbf{v}.\mathbf{A})\right]$$

$$= q\left[-\mathbf{\nabla}\phi - \frac{d\mathbf{A}}{dt} + \mathbf{\nabla}(\mathbf{v}.\mathbf{A})\right]. \tag{7.19}$$

Hence we can write

$$\frac{d}{dt}(m\mathbf{v} + q\mathbf{A}) = \mathbf{\nabla}(-q\phi + q\mathbf{v}.\mathbf{A}). \tag{7.20}$$

If equation (7.20) is to be put into the form (7.16), we must have a Lagrangian L defined such that, using Cartesian co-ordinates,

$$mv_j + qA_j = \frac{\partial L}{\partial \dot{x}_j}, \tag{7.21}$$

and with

$$\frac{\partial}{\partial x_j}(-q\phi + q\mathbf{v}.\mathbf{A}) = \frac{\partial L}{\partial x_j}. \tag{7.22}$$

Equations (7.21) and (7.22) are compatible with a Lagrangian L defined by

$$L = \tfrac{1}{2}mv^2 - q\phi + q\mathbf{v}.\mathbf{A}. \tag{7.23}$$

Hence given the electric potential ϕ and the magnetic vector potential $\mathbf{A}$ for any field we can construct the Lagrangian L from equation (7.23) and obtain the equations of motion for the particle.

We shall now consider an application of the Lagrangian method which is important in that the fields we consider are similar to those in certain electron devices, such as the magnetron valve.

▶ *Example*

Consider a pair of infinitely long cylinders, coaxial, of radii r_1 and r_2 with $r_2 > r_1$. Suppose a uniformly distributed current J flows along the inner cylinder. The cylinders are maintained at a potential difference V, the outer one being at the lower potential. There is an applied uniform magnetic field $\mathbf{B}$ parallel to the axis of the cylinders. Show that a charged particle of mass m and charge q which leaves the inner cylinder with negligible velocity will only just reach the outer cylinder if

$$V = \frac{q}{8m}\left\{\frac{\mu_0^2 J^2}{\pi^2}\left[\log\left(r_2/r_1\right)\right]^2 + \frac{B^2}{r_2^2}\left(r_2^2 - r_1^2\right)^2\right\}.$$

▶ *Solution*

We shall use a system of cylindrical polar co-ordinates with the axis of the cylinders as the z-axis, say (r, θ, z). Since the system is axially symmetric and invariant under motion in the z-direction, we must have an electric potential ϕ which is a function of r only. This is because the gradient of the potential gives the intensity, which must be radial. The Laplace equation for the potential becomes in this case

$$\frac{1}{r}\frac{d}{dr}\left(r\frac{d\phi}{dr}\right) = 0.$$

Integrating this equation once gives

$$r\frac{d\phi}{dr} = \text{constant} = C \text{ (say)},$$

and a second integration gives

$$\phi = C\log r + D,$$

where D is also a constant. If we now apply the boundary conditions for the potential,

$$\phi(r_1) = 0, \qquad \phi(r_2) = -V,$$

we obtain for the electric potential at a point distance r from the axis:

$$\phi(r) = \frac{-V\log(r/r_1)}{\log(r_2/r_1)}. \tag{1}$$

We have found the electric potential for the system; now in order to construct the Lagrangian we need the magnetic vector potential $\mathbf{A}$.

The magnetic intensity due to the current flowing along the inner cylinder, in the positive z direction, is

$$\mathbf{B}_1 = (0, \mu_0 J/2\pi r, 0). \tag{2}$$

The proof of this result is the same as that for the first example in Chapter 4.

The magnetic intensity of the applied field is

$$\mathbf{B}_2 = (0, 0, B). \tag{3}$$

The total magnetic intensity can now be written as

$$\mathbf{B} = (0, \mu_0 J/2\pi r, B). \tag{4}$$

A vector potential which leads to the field intensity given by equation (4) is

$$\mathbf{A} = [0, Br/2, -\mu_0 J\log(r/r_1)/2\pi]. \tag{5}$$

The Lagrangian for the motion of a charged particle in these fields can now be written as

$$L = \tfrac{1}{2}m(\dot{r}^2 + r^2\dot{\theta}^2 + \dot{z}^2) + \frac{qV \log (r/r_1)}{\log (r_2/r_1)} + q\left[\frac{Br^2\dot{\theta}}{2} - \frac{\mu_0 J\dot{z} \log (r/r_1)}{2\pi}\right].$$

(6)

We see that z is an ignorable co-ordinate, i.e. it does not appear in the Lagrangian, which implies that the conjugate momentum

$$p_z = \frac{\partial L}{\partial \dot{z}} = m\dot{z} - \frac{\mu_0 qJ \log (r/r_1)}{2\pi} = \text{constant}.$$

(7)

We can evaluate the constant from the initial condition that $\dot{z} = 0$ at $r = r_1$, to give

$$m\dot{z} - \frac{\mu_0 qJ \log (r/r_1)}{2\pi} = 0.$$

(8)

Since the Lagrangian does not contain the time explicitly, we have the Hamiltonian, which is the total energy of the particle, a constant. Hence, using equation (8) we have:

$$\frac{m}{2}(\dot{r}^2 + r^2\dot{\theta}^2) + \frac{m}{2}\frac{\mu_0^2 q^2 J^2}{4m^2\pi^2}[\log(r/r_1)]^2 - \frac{qV\log(r/r_1)}{\log(r_2/r_1)} = \text{constant}.$$

(9)

Using the initial conditions at $r = r_1$, we can put this constant equal to zero. Thus we have

$$\dot{r}^2 + r^2\dot{\theta}^2 + \frac{\mu_0^2 q^2 J^2}{4\pi^2 m^2}[\log (r/r_1)]^2 - \frac{2qV \log (r/r_1)}{m \log (r_2/r_1)} = 0.$$

(10)

The Lagrange equation of motion in θ is

$$\frac{d}{dt}\left(\frac{\partial L}{\partial \dot{\theta}}\right) - \frac{\partial L}{\partial \theta} = 0,$$

which implies that

$$mr^2\dot{\theta} + \tfrac{1}{2}qBr^2 = \text{constant}.$$

Using the initial conditions $\dot{\theta} = 0$ at $r = r_1$, we can put

$$\dot{\theta} = -qB(r^2 - r_1^2)/2mr^2.$$

(11)

From equations (10) and (11) we can deduce that

$$\dot{r}^2 = \frac{2qV \log (r/r_1)}{m \log (r_2/r_1)} - \frac{q^2 B^2 (r^2 - r_1^2)^2}{4m^2 r^2} - \frac{\mu_0^2 q^2 J^2 [\log (r/r_1)]^2}{4\pi^2 m^2}.$$

(12)

If the particle just reaches the outer cylinder we can put $\dot{r} = 0$ at $r = r_2$. This means that

$$V = \frac{q}{8m}\left(\frac{\mu_0^2 J^2[\log{(r_2/r_1)}]^2}{\pi^2} + \frac{B^2(r_2^2 - r_1^2)^2}{r_2^2}\right),$$

which is the required result.   ◀

▶ *Example*

Let us consider the example of an electron which is deflected from uniform motion by a uniform electric field. An electron of mass m and charge e is projected parallel to the x-axis from the origin with velocity **V**. There is a uniform electric field **E** parallel to the y-axis. Determine the equation of the orbit.

▶ *Solution*

Let us apply equation (7.4) to this problem. Since $\mathbf{r} = 0$ at $t = 0$, by a suitable choice of the zero of time, we can put $\mathbf{W} = 0$. Also the vector **V** is the initial velocity and is of magnitude V and parallel to the x-axis. The vector **E** in this case has magnitude E and is parallel to the y-axis. Let **i** and **j** be unit vectors along the x- and y-axes respectively; the equation of the orbit is

$$\mathbf{r}(t) = (eEt^2/2m)\mathbf{j} + (Vt)\mathbf{i}. \tag{1}$$

We can also write this equation of the orbit as

$$\mathbf{r}(t) = x\mathbf{i} + y\mathbf{j}, \tag{2}$$

where x and y are the co-ordinates of the particle at time t. Thus from equations (1) and (2) we have

$$x = Vt, \tag{3}$$

$$y = eEt^2/2m. \tag{4}$$

If we eliminate the time t from equations (3) and (4), we obtain the equation of the orbit in what is probably the best form

$$y = (eE/2mV^2)x^2, \tag{5}$$

which is a parabola.   ◀

This result can also be applied to the important problem of the deflection of a beam of electrons by an electric field, as in a cathode ray oscilloscope, provided one ignores the electromagnetic inter-actions between the electrons which make up the beam and treats each electron as if the others were absent. Inclusion of the inter-actions between the electrons leads to a discussion of the so called 'space charge' effects.

▶ *Example*

An electrode is at the origin of a system of cylindrical polar co-ordinates (r, θ, z). There is a screen in the plane $z = d$. There is a uniform magnetic field parallel to the z-axis. If an electron is emitted from the electrode with velocity $\mathbf{V}$ at an angle ψ to the z-axis, towards the screen, find the distance from the axis of the point at which the electron strikes the screen.

▶ *Solution*

Let e be the charge on the electron, m its mass and $\mathbf{B}$ the magnetic intensity. A vector potential which represents this magnetic field is

$$\mathbf{A} = \tfrac{1}{2}B(0, r, 0) \tag{1}$$

where we must remember that we are in cylindrical co-ordinates. There is no electric potential in this problem. The Lagrangian for the electron in this field is

$$L = \tfrac{1}{2}m(\dot{r}^2 + r^2\dot{\theta}^2 + \dot{z}^2) + eBr^2\dot{\theta}/2.$$

If we now introduce the Larmor* frequency ω defined by

$$\omega = eB/2m,$$

we can write the Lagrangian as

$$L = \tfrac{1}{2}m(\dot{r}^2 + r^2\dot{\theta} + \dot{z}^2) + m\omega r^2\dot{\theta}. \tag{2}$$

The Lagrange equation in the variable θ is

$$\frac{d}{dt}\left(\frac{\partial L}{\partial \dot{\theta}}\right) - \frac{\partial L}{\partial \theta} = \frac{d}{dt}(mr^2\dot{\theta} + mr^2\omega) = 0,$$

which integrates to give

$$r^2(\dot{\theta} + \omega) = \text{constant}.$$

---

* Sir J. Larmor (1857–1942), Irish mathematician.

7

Initially at the electrode we have $r = 0$, and so we can put this constant to zero, which implies that for all times

$$\dot{\theta} = -\omega. \tag{3}$$

Using this result we can now write the Lagrangian in the more compact form

$$L = \tfrac{1}{2}m(\dot{r}^2 + \dot{z}^2 - r^2\omega^2). \tag{4}$$

The Lagrange equation in the variable r gives us

$$\frac{d}{dt}\left(\frac{\partial L}{\partial \dot{r}}\right) - \frac{\partial L}{\partial r} = \ddot{r} + \omega^2 r = 0,$$

which on integration gives

$$\dot{r}^2 + \omega^2 r^2 = \text{constant.}$$

If we now use the initial condition that $\dot{r} = V \sin \psi$ at $r = 0$, we can evaluate the constant in the last equation to obtain

$$\dot{r}^2 + \omega^2 r^2 = V^2 \sin^2 \psi.$$

On further integration, and using the condition that the particle leaves the origin at time $t = 0$, we get

$$t = \int_0^r \frac{dn}{(V^2 \sin^2 \psi - \omega^2 n^2)^{1/2}} = \frac{\sin^{-1}(\omega r / V \sin \psi)}{\omega}, \tag{5}$$

which we can invert to give

$$r = V \sin \psi \sin(\omega t)/\omega. \tag{6}$$

The Lagrange equation in the variable z gives us

$$\frac{d}{dt}\left(\frac{\partial L}{\partial \dot{z}}\right) - \frac{\partial L}{\partial z} = m\ddot{z} = 0,$$

which integrates to give, using the initial conditions,

$$\dot{z} = V \cos \psi.$$

A second integration gives us

$$z = (V \cos \psi)t, \tag{7}$$

where we have used the fact that $z = 0$ at $t = 0$.

The distance from the z-axis of the point where the electron strikes the screen is obtained by calculating the time it takes the particle to reach the screen, which from (7) is equal to $(d/V \cos \psi)$, and inserting this into equation (6). The final result is

$$r = \frac{V \sin \psi \sin (\omega d/V \cos \psi)}{\omega}.$$ ◀

Problems for Chapter 7

(1) A charged particle is released from rest at a point in a region where there is a uniform electric field and a uniform magnetic field inclined at an angle. Show that the particle describes a cycloid in a plane which is perpendicular to the magnetic field and which moves parallel to it with constant acceleration.

(2) A particle of mass m and charge e at a distance D from a wire carrying a current J has a velocity $\mathbf{V}$ along an outward normal to the wire. Show that in the subsequent motion the distance of the particle from the wire varies between De^p and De^{-p}, where $p = 2\pi m V/e\mu_0 J$.

Chapter 8

FIELDS DUE TO MOVING CHARGED PARTICLES

8.1. The Fields of Moving Particles

In our discussion of electromagnetic fields, we have so far described the charge and current distributions by macroscopic density functions. These densities lead to results which represent the fields due to large numbers of charged particles. However, it often happens that practically as well as theoretically we are interested in phenomena which involve a knowledge of the field produced by a single charged particle. We find that to pass from our previous results to the new case requires a great deal of care.

We consider a charged fundamental particle to be well represented by a model in which a charge distribution is confined to a small region of space. By a small region of space we mean one whose maximum linear dimension is small compared to a characteristic dimension of the problem in hand. Each element of the distribution moves with the same velocity $\mathbf{v}$. Whether this is a good model or not depends on the agreement between the results we derive and observations.

The reason why we cannot immediately use our previously derived results, for example the expressions for the retarded potentials in equations (2.40) and (2.41), is that although when the charge distribution diminishes in dimension we could take the factor $|\mathbf{r} - \mathbf{r}'|^{-1}$ outside the integral sign, the integral

$$\int \int \int [\rho] \, d\tau$$

does not then represent the total charge of the system, since $[\rho]$ is evaluated for different times for different elements. We shall now consider the modifications that are needed to solve the problem.

8.2. THE LIÉNARD–WICHERT POTENTIALS

It is helpful if one can get a pictorial idea of the retarded potentials. The following idea is in many ways satisfactory (Fig. 9).

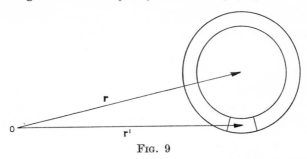

FIG. 9

Consider an observer at the point $\mathbf{r}$, where at time t he is interested in the electromagnetic field intensities. A sphere with centre $\mathbf{r}$ contracts with radial velocity c from a very large radius to reduce to the point $\mathbf{r}$ at time t. The time T at which the surface of the sphere passes the point $\mathbf{r}'$ is the time when the charge near this point makes its contribution to the retarded potential at $\mathbf{r}$ at time t. Here

$$T = t - |\mathbf{r} - \mathbf{r}'|/c. \tag{8.1}$$

We can imagine the spherical surface sweeping up the contributions to the potentials as it contracts. If the charge distribution is static, the amount of charge which an element dS of the surface at $\mathbf{r}'$ will cross in time dT during which the radius decreases by $d|\mathbf{r} - \mathbf{r}'|$ is

$$[\rho]\, dS\, d|\mathbf{r} - \mathbf{r}'|$$

where $[\rho]$ and dS are evaluated at T given by equation (8.1). If however the charge distribution is not static but at the point $\mathbf{r}'$ at time T has velocity $[\mathbf{v}]$, then a quantity of charge

$$[\rho]\, dT\, dS[\mathbf{v}.(\mathbf{r} - \mathbf{r}')]/[|\mathbf{r} - \mathbf{r}'|]$$

will flow from the element inwards before it can be met by the sphere. Again all quantities in square brackets are to be evaluated at the retarded time T. Thus the amount of charge de collected by the sphere as it passes over the element $d\tau = c\, dS\, dT$ will be

$$de = [\rho]\, d\tau - \frac{[\rho][\mathbf{v}.(\mathbf{r} - \mathbf{r}')]\, d\tau}{c[|\mathbf{r} - \mathbf{r}'|]}. \tag{8.2}$$

Hence

$$[\rho]\,d\tau = \frac{de}{1 - \dfrac{[\mathbf{v}\cdot(\mathbf{r}-\mathbf{r}')]}{c[|\mathbf{r}-\mathbf{r}'|]}} \qquad (8.3)$$

If we now substitute this expression into equations (2.40) and (2.41) and take the limit of a point charge we can put

$$\iiint de = e,$$

where e is the total charge on the system, since the contracting sphere sweeps over all the charge in the system. Then we obtain for the potentials of a point charge e and velocity $\mathbf{v}$, at the point $\mathbf{r}$ at time t,

$$\phi(\mathbf{r}, t) = \frac{e}{4\pi\epsilon_0\left\{[|\mathbf{r}-\mathbf{r}'|] - \dfrac{[\mathbf{v}\cdot(\mathbf{r}-\mathbf{r}')]}{c}\right\}}, \qquad (8.4)$$

$$\mathbf{A}(\mathbf{r}, t) = \frac{\mu_0 e[\mathbf{v}]}{4\pi\left\{[|\mathbf{r}-\mathbf{r}'|] - \dfrac{[\mathbf{v}\cdot(\mathbf{r}-\mathbf{r}')]}{c}\right\}}. \qquad (8.5)$$

These expressions are known as the Liénard–Wichert potentials.* They refer to a point charge, or at least to a charge with small dimension, but are otherwise independent of the details of the particle model.

Although these potentials look fairly simple, this is deceptive, since they contain the velocity of the particle. This velocity can only be determined provided we know the forces acting on the particle and hence the potentials. Thus we are in a vicious circle. To know the field we need the velocity for which in turn we need the field. To get out of this dilemma we normally proceed as follows. We assume that there are forces acting on the charged particle which produce a motion on which the effect of the field produced by the particle itself is negligible. Thus by using only external forces in the equation of motion we can calculate the position of the particle $\mathbf{r}'(t)$ at any time t to a high degree of accuracy. If we now want to calculate the field produced by the particle at the point $\mathbf{r}$ at time t, we take the function $\mathbf{r}'$ we have just calculated and from the equation

$$T = t - |\mathbf{r}-\mathbf{r}'(T)|/c \qquad (8.6)$$

* A. Liénard (1869–   ); E. Wichert (1861–1928), German geophysicist.

we calculate the retarded time T. We then evaluate $[\mathbf{v}] \cdot [|\mathbf{r}-\mathbf{r}'|]$ and $[\mathbf{v} \cdot (\mathbf{r}-\mathbf{r}')]$ at this retarded time, and insert them into equations (8.4) and (8.5) to give the required potentials.

It is implicit in our work that the charge on a particle is an absolute constant and not dependent on the motion of the charge. Our belief in the validity of this assumption is again due to the agreement between the theory based on it and experimental results.

8.3. The Field Intensities due to Moving Charged Particles

We shall now turn to the problem of deriving the intensities of the electric and magnetic fields due to a moving charged particle from the Liénard–Wichert potentials. These intensities are the basic fields of classical electrodynamics, and are of great importance.

At a point $\mathbf{r}$ at time t we derive the field intensities by differentiating the potentials with respect to $\mathbf{r}$ and t, as in equation (7.20). However the actual information about the motion of the charged particle is given as functions of the retarded time T, given by equation (8.6). We must therefore relate derivatives with respect to $\mathbf{r}$ and t to the derivatives with respect to T. This will be our first problem.

First we have

$$\frac{\partial[|\mathbf{r}-\mathbf{r}'|]}{\partial t} = \frac{1}{2[|\mathbf{r}-\mathbf{r}'|]} \frac{\partial}{\partial T} [(\mathbf{r}-\mathbf{r}') \cdot (\mathbf{r}-\mathbf{r}')] \frac{\partial T}{\partial t}$$

$$= -\left[\frac{\mathbf{v} \cdot (\mathbf{r}-\mathbf{r}')}{|\mathbf{r}-\mathbf{r}'|}\right] \frac{\partial T}{\partial t}. \tag{8.7}$$

From equation (8.6) however we obtain

$$\frac{\partial T}{\partial t} = 1 - \frac{1}{c} \frac{\partial}{\partial t} [|\mathbf{r}-\mathbf{r}'|]. \tag{8.8}$$

We now combine equations (8.7) and (8.8) to give

$$\frac{\partial T}{\partial t} = \left\{1 - \frac{[\mathbf{v} \cdot (\mathbf{r}-\mathbf{r}')]}{c[|\mathbf{r}-\mathbf{r}'|]}\right\}^{-1}. \tag{8.9}$$

In order to make some of the calculations easier, we introduce the new variable S defined by

$$S = [|\mathbf{r}-\mathbf{r}'|] - [\mathbf{v} \cdot (\mathbf{r}-\mathbf{r}')]/c. \tag{8.10}$$

In this new notation, equation (8.9) becomes

$$\frac{\partial T}{\partial t} = \frac{[|\mathbf{r}-\mathbf{r}'|]}{S}. \tag{8.11}$$

Thus we have obtained one relation between variables at the field point $(\mathbf{r}, t)$ and the source point $(\mathbf{r}', T)$.

Now consider equation (8.6) again. We have

$$\mathbf{\nabla}_{\mathbf{r}}[|\mathbf{r}-\mathbf{r}'|] = -c\mathbf{\nabla}_{\mathbf{r}}T \tag{8.12}$$

where the gradient operators contain derivatives with respect to the field point $\mathbf{r}$, as indicated by the subscript. But we can also write

$$\mathbf{\nabla}_{\mathbf{r}}[|\mathbf{r}-\mathbf{r}'|] = \frac{[\mathbf{r}-\mathbf{r}']}{[|\mathbf{r}-\mathbf{r}'|]} + \frac{\partial}{\partial T}[|\mathbf{r}-\mathbf{r}'|]\mathbf{\nabla}_{\mathbf{r}}T, \tag{8.13}$$

since $[|\mathbf{r}-\mathbf{r}'|]$ is a function of $\mathbf{r}$ explicitly and T, which is also a function of $\mathbf{r}$. From equations (8.12), (8.13), (8.10), (8.11) and (8.7) we obtain

$$\mathbf{\nabla}_{\mathbf{r}}T = -\frac{[\mathbf{r}-\mathbf{r}']}{[|\mathbf{r}-\mathbf{r}'|]}\left\{c+\frac{\partial[|\mathbf{r}-\mathbf{r}'|]}{\partial T}\right\}^{-1} = -\frac{[\mathbf{r}-\mathbf{r}']}{cS}, \tag{8.14}$$

giving a second relationship between derivatives at $(\mathbf{r}, t)$ and $(\mathbf{r}', T)$.

Using equations (8.14) and (8.11), we can now write the gradient of the electric potential, which occurs in the electric intensity, as a function of the variables at the retarded time. We obtain from equations (8.4) and (8.10)

$$\mathbf{\nabla}_{\mathbf{r}}\phi = -\frac{e}{4\pi\epsilon_0 S^2}\mathbf{\nabla}_{\mathbf{r}}S$$

$$= -\frac{e}{4\pi\epsilon_0 S^2}\left\{\mathbf{\nabla}_{\mathbf{r}}[|\mathbf{r}-\mathbf{r}'|]-\frac{\mathbf{\nabla}_{\mathbf{r}}[\mathbf{v}\cdot(\mathbf{r}-\mathbf{r}')]}{c}\right\}, \tag{8.15}$$

which from equation (8.12) can be written

$$\mathbf{\nabla}_{\mathbf{r}}\phi = \frac{e}{4\pi\epsilon_0 S^2}\left\{c\mathbf{\nabla}_{\mathbf{r}}T+\frac{1}{c}\mathbf{\nabla}_{\mathbf{r}}[\mathbf{v}\cdot(\mathbf{r}-\mathbf{r}')]\right\}. \tag{8.16}$$

Now

$$\mathbf{\nabla}_{\mathbf{r}}[\mathbf{v}\cdot(\mathbf{r}-\mathbf{r}')] = [\mathbf{v}]+\frac{\partial}{\partial T}[\mathbf{v}\cdot(\mathbf{r}-\mathbf{r}')]\mathbf{\nabla}_{\mathbf{r}}T$$

$$= [\mathbf{v}]-[\mathbf{v}^2]\mathbf{\nabla}_{\mathbf{r}}T+[\dot{\mathbf{v}}\cdot(\mathbf{r}-\mathbf{r}')]\mathbf{\nabla}_{\mathbf{r}}T. \tag{8.17}$$

From equations (8.17), (8.16) and (8.15) we finally obtain

$$\mathbf{V}_r\phi = \frac{e}{4\pi\epsilon_0 S^2}\left\{\frac{[\mathbf{r}-\mathbf{r}']}{S}\left(\frac{[v^2]}{c^2}-\frac{[\dot{\mathbf{v}}\cdot(\mathbf{r}-\mathbf{r}')]}{c^2}-1\right)+\frac{[\mathbf{v}]}{c}\right\}. \quad (8.18)$$

We have calculated one term of the electric intensity; now let us turn to the other. From equations (8.5) and (8.10) we have

$$\frac{\partial\mathbf{A}}{\partial t} = \frac{\partial}{\partial t}\left(\frac{\mu_0 e[\mathbf{v}]}{4\pi S}\right) = \frac{\mu_0}{4\pi}\left\{\frac{e}{S}\frac{\partial[\mathbf{v}]}{\partial T}\frac{\partial T}{\partial t}-\frac{e[\mathbf{v}]}{S^2}\frac{\partial S}{\partial t}\right\},$$

which on using equation (8.11) becomes

$$\frac{\partial\mathbf{A}}{\partial t} = \frac{\mu_0 e}{4\pi S^2}\left\{[\dot{\mathbf{v}}][|\mathbf{r}-\mathbf{r}'|]-[\mathbf{v}]\frac{\partial}{\partial t}\left([|\mathbf{r}-\mathbf{r}'|]-\frac{[\mathbf{v}\cdot(\mathbf{r}-\mathbf{r}')]}{c}\right)\right\}. \quad (8.19)$$

Substituting from equations (8.7) and (8.11) into equation (8.19), we obtain:

$$\frac{\partial\mathbf{A}}{\partial t} = \frac{\mu_0 e}{4\pi S^2}\left\{[\dot{\mathbf{v}}][|\mathbf{r}-\mathbf{r}'|]\right.$$
$$\left.-[\mathbf{v}]\left(c\left(1-\frac{[|\mathbf{r}-\mathbf{r}'|]}{S}\right)-\frac{[|\mathbf{r}-\mathbf{r}'|]}{cS}[\dot{\mathbf{v}}\cdot(\mathbf{r}-\mathbf{r}')-v^2]\right)\right\}. \quad (8.20)$$

We now have the other term in the electric intensity. Combining equations (8.20) with (8.17) into

$$\mathbf{E} = -\mathbf{V}_r\phi - \frac{\partial\mathbf{A}}{\partial t},$$

we finally obtain

$$4\pi\epsilon_0\mathbf{E} = \frac{-e}{S^2}\left\{\frac{[\mathbf{r}-\mathbf{r}']}{S}\left(\frac{[v^2]}{c^2}-\frac{[\dot{\mathbf{v}}\cdot(\mathbf{r}-\mathbf{r}')]}{c^2}-1\right)\right\}$$
$$-\frac{e[|\mathbf{r}-\mathbf{r}'|]}{cS^2}\left\{\frac{[\dot{\mathbf{v}}]}{c}+\frac{[\mathbf{v}][\dot{\mathbf{v}}\cdot(\mathbf{r}-\mathbf{r}')]}{c^2 S}+\frac{[\mathbf{v}]}{S}\left(1-\frac{[v^2]}{c^2}\right)\right\}. \quad (8.21)$$

We obtain the magnetic intensity from the relation

$$\mathbf{B} = \mathbf{V}_r\times\mathbf{A} = \mathbf{V}_r\times\left(\frac{\mu_0 e[\mathbf{v}]}{4\pi S}\right)$$

or

$$4\pi\mathbf{B} = \frac{\mu_0 e}{S}\left\{\mathbf{V}_r\times[\mathbf{v}]+\frac{[\mathbf{v}]}{S}\times\mathbf{V}_r S\right\}$$
$$= \frac{\mu_0 e}{S}\left\{\mathbf{V}_r T\times[\dot{\mathbf{v}}]+\frac{[\mathbf{v}]}{S}\times\mathbf{V}_r S\right\}, \quad (8.22)$$

We calculated $\nabla_r S$ implicitly when we went from (8.15) to (8.18), so we can write

$$4\pi\mathbf{B} = \frac{\mu_0 e}{S}\left\{-\frac{[\mathbf{r}-\mathbf{r}']\times[\dot{\mathbf{v}}]}{cS}+\frac{[\mathbf{r}-\mathbf{r}']}{S^2}\left(\frac{[v^2]}{c^2}-\frac{[\dot{\mathbf{v}}\cdot(\mathbf{r}-\mathbf{r}')]}{c^2}-1\right)\times[\mathbf{v}]\right\}$$

which reduces to

$$\mathbf{B} = \frac{[\mathbf{r}-\mathbf{r}']\times\mathbf{E}}{[|\mathbf{r}-\mathbf{r}'|]c} \tag{8.23}$$

Thus at long last we have obtained the electric and magnetic intensities of the field due to a point charge in an arbitrary state of motion. We have based our discussion solely on the retarded fields. These fields which we have derived are of fundamental importance, since any electromagnetic field must ultimately be a superposition of such fields.

Since the field intensities we have calculated are so fundamental, it is disturbing that they show so clearly the weaknesses of classical electromagnetic theory. Just as we can only derive potentials which are useful when we neglect the effect of the particle's own field on its motion, these intensities suffer from the same defect.

If the flux of the Poynting vector derived from these fields is taken over a large sphere which contains an accelerating particle near its centre, it is found to be almost independent of the radius of the sphere. This suggests that energy is continuously radiated in the form of electromagnetic waves by the particle. This radiated energy must be accompanied by a loss in the kinetic energy of the particle, but the exact procedure for relating the two is not yet known. These are some of the outstanding problems of electrodynamics. As mentioned earlier in this book, some of them have been tackled using advanced potentials, but this does not seem to have removed all the difficulties; in fact it introduces new ones.

Now let us consider an example to illustrate the results we have derived, and to establish the much-used Larmor formula for the energy radiated by a slow-moving electron.

▶ *Example*

That component of the electromagnetic field of a charged particle which depends on the acceleration is often termed the *radiation field*, since it decreases more slowly with distance from the particle than the other components, and may thus be thought of as carrying energy further away from the particle. Let us determine the radiation field for a particle whose velocity is small compared to the velocity of light.

▶ *Solution*

Let **v** be the velocity of the particle. Then $v \ll c$, and we shall only retain the leading terms of the field intensities expanded in terms of the parameter (v/c).

If from equations (8.21) and (8.23) we extract the terms which depend on the acceleration, we obtain:

$$E = \frac{-e}{4\pi c^2 S^2 \epsilon_0} \left\{ [\dot{\mathbf{v}}][\mathbf{R}] - \frac{[\dot{\mathbf{v}} \cdot \mathbf{R}][\mathbf{R}]}{S} \right\}, \tag{1}$$

$$B = \frac{\mu_0 e [\dot{\mathbf{v}} \times \mathbf{R}]}{4\pi c S^2}, \tag{2}$$

where $\mathbf{R} = \mathbf{r} - \mathbf{r}'$.

We can simplify these expressions for the intensities further if we note that from equation (8.10) we can to first order put

$$S = [R]. \tag{3}$$

Then equations (1) and (2) become

$$E = \frac{e[\mathbf{R} \times (\mathbf{R} \times \dot{\mathbf{v}})]}{4\pi\epsilon_0 c^2 [R]^3}, \tag{4}$$

$$B = \frac{\mu_0 e [\dot{\mathbf{v}} \times \mathbf{R}]}{4\pi c [R]^2}. \tag{5}$$

Since the retarded values differ from the present values by terms of order (v/c), we can to first order drop the square brackets which denote retarded values and let all quantities refer to the instant of measurement.

Let us now consider how this radiation field affects the energy of the charged particle. We must calculate the Poynting vector $(\mathbf{E} \times \mathbf{B})/\mu_0$ corresponding to this field and determine its flux over some surface which encloses the particle. This will give us the rate at which energy is leaving this region due to the radiation field, and we must assume that this energy is derived from the kinetic energy of the motion of the charge. Note, however, that we are only considering the radiation field and not the *total* field; hence we must not expect our results to be accurate in detail—a point we shall return to later.

For convenience of calculation we take as our surface of integration a spherical surface with centre at the instantaneous position of the particle. First we have:

$$\mathbf{E} \times \mathbf{B} = \frac{\mu_0 e^2[(\mathbf{R} \cdot \dot{\mathbf{v}})\mathbf{R} \times (\dot{\mathbf{v}} \times \mathbf{R}) - R^2\dot{\mathbf{v}} \times (\dot{\mathbf{v}} \times \mathbf{R})]}{16\pi^2 c^3 \epsilon_0 R^5}$$

$$= \frac{\mu_0 e^2[R^2\dot{\mathbf{v}}^2 - (\mathbf{R} \cdot \dot{\mathbf{v}})^2]\mathbf{R}}{16\pi^2 \epsilon_0 c^3 R^5}. \tag{6}$$

Let θ be the angle between $\dot{\mathbf{v}}$ and $\mathbf{R}$. Then for the flux of the Poynting vector over the spherical surface S we obtain:

$$\frac{1}{\mu_0} \iint_S (\mathbf{E} \times \mathbf{B}) \cdot d\mathbf{S} = \frac{e^2\dot{v}^2}{16\pi^2\epsilon_0 c^3} \int_0^\pi (1 - \cos^2\theta)2\pi \sin\theta \, d\theta$$

$$= \frac{\mu_0 e^2\dot{v}^2}{6\pi c^3}. \tag{7}$$

This result for the 'radiated power' from a nonrelativistic charged particle is known as *Larmor's formula*. It plays an important part in discussions of the effect of the self-field of a particle on its motion. ◀

CHAPTER 9

THE PROBLEM OF CHARGED PARTICLES
IN CLASSICAL ELECTROMAGNETIC THEORY

9.1. THE SELF-FIELD

In the theory of the electromagnetic field which we have developed so far in this work, there are great difficulties in treating the problems of the structure and motion of charged particles. In the classical theory of the electromagnetic field, the charged particle is thought of as an entity which is distinct from the field and, the theory being linear in the sense we have discussed earlier, the field acting on any charged particle is normally considered in two parts. One part is said to be due to the particular charged particle under consideration, and the other to an 'external' field, although of course this also is ultimately due to other charged particles. It is that part of the field associated with the charged particle itself, the so-called 'self-field', which is the great problem in the theory; it has many undesirable features which suggest that either the way in which we form models of charged particles is at fault, or else the treatment of electromagnetic fields which we have developed needs to be modified.

9.2. ENERGY OF THE SELF-FIELD

If we consider ourselves in a frame of reference in which we have a charged particle at rest, then either we can consider the particle to be a true point charge of zero spatial extent, or we can consider it to be a distribution of charge and mass within a volume which is small on the macroscopic scale. One of the simplest and most plausible of the latter type of model is a uniformly charged sphere of definite radius, R (say). We shall now consider some of the problems associated with each of these models.

Let us consider first the model of an electron at rest as a true point charge of zero spatial extent. We have shown on the basis of the

97

accepted field equations that at a distance r from the electron, which we shall suppose to carry a charge e, the electric intensity $\mathbf{E}$ of the self-field has magnitude $e/4\pi\epsilon_0 r^2$. The magnetic intensity of the self-field is zero at all points. This means that at each point of space we have an energy density distribution associated with this self-field which is given by Poynting's equation as

$$\epsilon_0 E^2/2 = e^2/32\epsilon_0\pi^2 r^4. \tag{9.1}$$

To obtain the total energy residing in the field of the electron, we have to integrate the expression (9.1) over the whole of space. Here lies the undesirable feature. Denote this total energy by W and use spherical polar co-ordinates (r, θ, ϕ) with the electron as origin. Then

$$W = \int\limits_0^{2\pi} \int\limits_0^{\pi} \int\limits_0^{\infty} (e^2/32\epsilon_0\pi^2 r^4) r^2 \sin\theta \, dr \, d\theta \, d\phi$$

$$= \frac{e^2}{8\pi\epsilon_0} \int\limits_0^{\infty} \frac{dr}{r^2},$$

which diverges. Hence we see that associated with the self-field of a point electron we have an infinite amount of energy.

If we also bear in mind the mass-energy relation of Einstein * we see that the region occupied by the field within any sphere of nonzero radius and with centre at the electron has infinite energy content and hence infinite mass. We know that the self-field of an electron is carried along with the electron as it moves, and if the above discussion is correct then we would expect the electron always to exhibit an infinite inertial mass. This seems physically unreasonable and, even more important, is in direct contradiction to the well-established fact that a finite force can produce a finite acceleration of an electron.

We see that the idea of a point electron, or other charged particle, is very unsatisfactory. The infinite mass of the self-field cannot be compensated by that of other electromagnetic fields, since the mass of electromagnetic fields is essentially positive. Attempts have been made, notably by Dirac,† to compensate by an infinite negative gravitational mass. In this case the difference between the infinite

---

* A. Einstein (1879–1955).

† P. A. M. Dirac (1902–   ), British physicist; one of the pioneers of quantum mechanics.

positive mass of the self-field and the infinite negative gravitational mass is taken to be the observed mass of the electron. However, this point of view has its own inherent difficulties.

If we now try to circumvent the difficulties of the point electron by taking an alternative model of a charge distribution of finite extent, then we are again faced with difficulties. Since like charges repel each other, it is impossible to have a distribution of like charges in equilibrium under its own electric interaction alone; one must introduce additional forces to keep the system in equilibrium. These additional forces were first discussed by Poincaré*: they are often referred to as *Poincaré stresses*. The unsatisfactory nature of these stresses is that they are introduced into the theory *ad hoc* to explain one particular phenomenon, the stability of the charged particle, and seem to give no other evidence of, or reason for, their existence in the model. There are also difficulties with the stresses in regard to their transformation properties as one changes from one inertial frame of reference to another, but we shall not deal with this point here.

It seems as if both point and extended models of the charged particle have defects. One can try a different model. In this, one alters the equations of the electromagnetic field in such a way that at distances away from the particle greater than a certain critical distance, they agree with Maxwell's field equations; for distances less than this they have a behaviour which removes the divergence from the mass–energy integral. This type of approach is mainly the work of Born† and Infeld,‡ but has not been very successful.

9.3. The Classical Radius of the Electron

From the foregoing discussion one can obtain a parameter, which is called the *classical radius of the electron*. This is used a great deal in discussions on the structure and behaviour of charged particles, and occasionally in cosmological speculations.

Suppose we consider an electron of charge e as a spherically

---

* H. Poincaré (1854–1912), distinguished French mathematician and philosopher of science.

† M. Born (1882–   ), German physicist. He is one of the originators of modern quantum mechanics.

‡ L. Infeld (1898–   ), Polish physicist. He has worked mainly on problems in general relativity.

symmetric charge distribution within a sphere of radius R. What is
the value which this radius R must have so that the electrostatic
energy of the field outside the sphere is equivalent to the rest energy
of the electron? To obtain R we use a revised form of equation (9.2)
and put

$$mc^2 = \frac{e^2}{8\pi\epsilon_0} \int_R^\infty \frac{dr}{r^2} = \frac{e^2}{8\pi\epsilon_0 R}. \tag{9.3}$$

Hence we obtain for the classical radius of the electron

$$R = e^2/8\pi\epsilon_0 c^2 m. \tag{9.4}$$

Naturally this quantity has the dimensions of length.

It now seems that, apart from giving an idea of the order of
magnitude of the lengths one meets in problems of electron theory,
the *precise* value of the classical radius of the electron is of no particular significance.

9.4. Outline of the Dynamical Problem

Let us now very briefly consider the problems involved in determining the motion of a charged particle in an electromagnetic
field, or indeed in any force field. We see that expressions (8.21) and
(8.23) for the self-field of a moving charged particle are singular at
the particle itself, remembering that we are using the fully retarded
solutions, which means that we cannot hope to insert these fields
into the equation of motion

$$m\ddot{\mathbf{r}} = e[\mathbf{E} + (\dot{\mathbf{r}} \times \mathbf{B})] \tag{9.5}$$

and obtain a solution. We have to try and obtain the motion of the
particle by a more indirect method. This problem has never been
really satisfactorily solved, but one can obtain some information by
the following procedure (we give no detailed calculations but merely
outline the manner in which to proceed).

From the fields (8.21) and (8.23) we can obtain the Poynting
vector $\mathbf{P}$ for an electron moving in an arbitrary manner. Let the
electron be at the point $\mathbf{r}$ at time t. We can obtain the flux of $\mathbf{P}$
out of the surface S of a large sphere with centre $\mathbf{r}$ and radius R at a
time $t + R/c$. We then assume that this energy loss from the sphere is
made at the expense of the kinetic energy of the particle, and that it

can be put in the form of a force acting on the particle at time t. This force $\mathbf{X}$ is called the *radiation reaction*, and the rate at which it does work on the particle at time t at position $\mathbf{r}$ is the rate at which energy flows from the sphere S at time $t + R/c$.

The next step is to write the equation of motion for the electron or other charged particle as

$$m\ddot{\mathbf{r}} = e[\mathbf{E}' + (\dot{\mathbf{r}} \times \mathbf{B}')] + \mathbf{X}, \qquad (9.6)$$

where the primed fields $\mathbf{E}'$ and $\mathbf{B}'$ denote fields other than the self-field of the particle whose motion we are interested in. We then solve equation (9.6) for the position $\mathbf{r}(t)$ of the electron. This when inserted into equations (8.21) and (8.23) can give us the field of the moving electron.

The whole of the procedure we have outlined above is a very rough method of accounting for the effect of the self-field of a charged particle on its motion and is riddled with pitfalls and unsatisfactory assumptions. At best it could lead to an equation which represented a sort of average motion of the electron; any hope of detail is lost. The dynamical problem is so difficult that a really satisfactory treatment has not yet been given.

CONCLUSION

The material discussed in this chapter gives some indication of the problems which occupy the attention of those engaged in studying the fundamentals of electromagnetic theory at the present time. The problems cannot be avoided by using quantum electrodynamics; in many instances they actually become worse in the quantum theory, which may be a reason for trying to make progress first of all in the classical theory.

Appendix I

SELECT BIBLIOGRAPHY

The reader should find the following books useful as references for the subjects indicated.

VECTOR ANALYSIS
Introduction to Vector Analysis, H. F. Davis. Prentice-Hall (1961)
POTENTIAL THEORY
Solutions of Laplace's Equation, D. R. Bland. Routledge & Kegan Paul (1961)
FOURIER ANALYSIS
Fourier Series, G. P. Tolstov. Prentice-Hall (1962)
Fourier Transforms, I. N. Sneddon. McGraw-Hill (1951)

For more comprehensive books on electromagnetic theory, the reader is referred to the following.

Classical Electrodynamics, J. D. Jackson. Wiley (1962)
Classical Electricity and Magnetism, W. K. H. Panofsky and M. Philips. Addison-Wesley (1962)
The Feynman Lectures on Physics, Vol. 2, Feynman, Leighton and Sands. Addison-Wesley (1964)

For more detailed discussion of the contents of Chapter 9, the following book is excellent.

Classical Charged Particles, F. Rohrlich. Addison-Wesley (1965)

Appendix II

FUNDAMENTAL CONSTANTS

Charge on an electron $= 1{\cdot}6 \times 10^{-19}$ coulombs
Mass of an electron $= 9{\cdot}1 \times 10^{-31}$ kilograms
Mass of a proton $= 1{\cdot}67 \times 10^{-27}$ kilograms
$\mu_0 = 4\pi \times 10^{-7}$ henrys/metre
$\epsilon_0 = 8{\cdot}854 \times 10^{-12}$ farads/metre
$c = 2{\cdot}997 \times 10^8$ metres/second

SOLUTIONS TO PROBLEMS

CHAPTER 1

Problem (1)

The force of attraction between the charged bodies is radial and has a magnitude $-ee'/4\pi\epsilon_0 r^2$. The acceleration of the particle with mass m and charge e' in the radial direction due to its circular motion is v^2/r, where $\mathbf{v}$ is its velocity. Equating force to mass times acceleration, we obtain

$$-ee'/4\pi\epsilon_0 r^2 = mv^2/r.$$

Hence

$$v^2 = -ee'/4\pi\epsilon_0 mr.$$

Problem (2)

Take the divergence of both sides of the field equation

$$\nabla \times \mathbf{E} = -\frac{\partial \mathbf{B}}{\partial t}$$

to obtain

$$\nabla . (\nabla \times \mathbf{E}) = -\nabla . \frac{\partial \mathbf{B}}{\partial t}. \tag{1}$$

But $\nabla . (\nabla \times \mathbf{E}) = 0$, and if we assume that we can invert the order of the differentiation, we obtain from (1):

$$\frac{\partial(\nabla . \mathbf{B})}{\partial t} = 0. \tag{2}$$

If at any instant $\mathbf{B}$ is everywhere zero, then so is $\nabla . \mathbf{B}$. Equation (2) ensures that it remains so at future times.

Problem (3)

See Section 7.1.

CHAPTER 2

Problem (1)

The equation of continuity of charge is

$$\frac{\partial \rho}{\partial t} + \mathbf{\nabla} . \mathbf{j} = 0. \tag{1}$$

We also have the field equation

$$\mathbf{\nabla} . \mathbf{E} = \rho/\epsilon_0 \tag{2}$$

and

$$\mathbf{j} = \sigma \mathbf{E}. \tag{3}$$

If we take the divergence of equation (3) and use equation (2), we get

$$\mathbf{\nabla} . \mathbf{j} = \sigma \mathbf{\nabla} . \mathbf{E} = \sigma \rho/\epsilon_0. \tag{4}$$

From equations (4) and (1) we obtain

$$\frac{\partial \rho}{\partial t} + \frac{\sigma}{\epsilon_0} \rho = 0. \tag{5}$$

The solution for $\rho(t)$ is

$$\rho(\mathbf{r}, t) = \rho(\mathbf{r}, 0) \exp\left[(-\sigma/\epsilon_0)t\right],$$

and since $\sigma > 0$ this shows the exponential decrease which we set out to find.

Problem (2)

Take the divergence of the field equation

$$\mathbf{\nabla} \times \mathbf{B} = \mu_0\left(\mathbf{j} + \epsilon_0 \frac{\partial \mathbf{E}}{\partial t}\right). \tag{1}$$

This gives

$$\mathbf{\nabla} . \mathbf{j} + \epsilon_0 \mathbf{\nabla} . \frac{\partial \mathbf{E}}{\partial t} = 0. \tag{2}$$

But from the equation of continuity of charge, we have

$$\mathbf{\nabla} . \mathbf{j} = -\frac{\partial \rho}{\partial t}. \tag{3}$$

If we insert equation (3) into equation (2), we obtain

$$\frac{\partial \rho}{\partial t} = \epsilon_0 \mathbf{\nabla} \cdot \frac{\partial \mathbf{E}}{\partial t},$$

and on inverting the order of differentiation on the right-hand side,

$$\frac{\partial}{\partial t}(\epsilon_0 \mathbf{\nabla} \cdot \mathbf{E} - \rho) = 0. \tag{4}$$

If at any instant $\mathbf{E}$ and ρ are zero, then equation (4) implies that at any subsequent time we can satisfy the field equation

$$\mathbf{\nabla} \cdot \mathbf{E} = \rho/\epsilon_0.$$

Problem (3)

In a region free from current, the field equation

$$\mathbf{\nabla} \times \mathbf{B} = \frac{1}{c^2} \frac{\partial \mathbf{E}}{\partial t} \tag{1}$$

suggests that we put

$$\mathbf{E} = \mathbf{\nabla} \times \mathbf{T}. \tag{2}$$

If we invert the order of differentiation, equation (1) then becomes

$$\mathbf{\nabla} \times \left(\mathbf{B} - \frac{1}{c^2} \frac{\partial \mathbf{T}}{\partial t}\right) = 0. \tag{3}$$

Hence we can put

$$\mathbf{B} - \frac{1}{c^2} \frac{\partial \mathbf{T}}{\partial t} = -\mathbf{\nabla}\psi \tag{4}$$

for some scalar ψ. Thus we have the required field representation

$$\mathbf{E} = \mathbf{\nabla} \times \mathbf{T}, \qquad \mathbf{B} = -\mathbf{\nabla}\psi + \frac{1}{c^2} \frac{\partial \mathbf{T}}{\partial t}.$$

$$\mathbf{\nabla} \cdot \mathbf{E} = \mathbf{\nabla} \cdot (\mathbf{\nabla} \times \mathbf{T}) = 0,$$

which implies that we must have $\rho = 0$.

Problem (4)

The problem of obtaining the advanced potentials is almost identical with that of obtaining the retarded potentials: the same formal analysis used in Section 2.5 gives the results.

CHAPTER 3

Problem (*1*)

The system is axially symmetric about the line containing the charge. Hence the electric intensity at any point must be independent of the angular co-ordinate θ. The system is also invariant under a translation parallel to the axis of the cylinder, and hence the electric intensity must also be independent of z. The system is symmetric with respect to a plane normal to the line of charge, and hence the electric intensity must lie in such planes. Since the intensity at a point due to each element of charge lies in a plane which contains that point and the line of charge, the total intensity must also lie in this plane. Hence, by symmetry considerations, we have shown that at any point the electric intensity is along an outward normal from the line of charge, and that the electric intensity is independent of θ and z.

Consider the flux

$$F = \iint_S \mathbf{E} \cdot \mathbf{dS}$$

of the electric intensity over a surface S formed by a cylinder of radius r, with axis on the line of charge, and two planes normal to the line of charge and distance L apart (see Fig. 10). The contribution

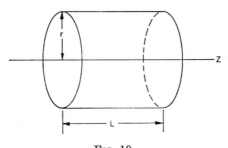

FIG. 10

to the flux from the flat ends of the cylinder is zero, since the electric intensity is radial to the line of charge. Over the curved surface $\mathbf{E}$ is normal and constant; hence

$$F = 2\pi r \cdot L \cdot E_r.$$

However, by Gauss' Theorem

$$F = \text{(total charge enclosed by } S)/\epsilon_0$$
$$= Lq/\epsilon_0.$$

Hence we must have

$$2\pi r L E_r = Lq/\epsilon_0,$$

which gives

$$E_r = q/2\pi\epsilon_0 r.$$

The potential which gives this intensity as its gradient is

$$\phi(r) = \alpha - (q/2\pi\epsilon_0) \log r,$$

where α is a constant.

Problem (2)

We have to solve Laplace's equation

$$\frac{\partial^2\phi}{\partial x^2} + \frac{\partial^2\phi}{\partial y^2} + \frac{\partial^2\phi}{\partial z^2} = 0 \tag{1}$$

for the region $0 \leqslant x \leqslant a$, $0 \leqslant y \leqslant b$, $0 \leqslant z \leqslant c$. The boundary conditions are:

$$\phi = 0 \quad \text{on} \quad x = 0, \, x = a, \, y = 0, \, y = b, \, z = 0, \tag{2}$$

$$= xy(x-a)(y-b) \quad \text{on} \quad z = c. \tag{3}$$

As we did in the worked example, we now choose a solution of equation (1) having the form

$$\phi(x, y, z) = \sum_{m,n} P_{mn} \sin (m\pi x/a) \sin (n\pi y/b) \sinh (b_{mn}z), \tag{4}$$

where the P_{mn} are constants and

$$b_{mn} = \pi(m^2/a^2 + n^2/b^2)^{1/2}. \tag{5}$$

This form of solution automatically satisfies conditions (2).

Now we express the function

$$f(x, y) = xy(x-a)(y-b),$$

which appears in condition (3), as a double Fourier series:

$$f(x, y) = \sum_{m \, n} C_{mn} \sin (m\pi x/a) \sin (n\pi y/b) \tag{6}$$

where the coefficients C_{mn} are given by

$$C_{mn} = \frac{4}{ab} \int_0^a \int_0^b f(x, y) \sin (m\pi x/a) \sin (n\pi y/b)\, dx\, dy$$

$$= \frac{4}{ab} \left(\int_0^a x(x-a) \sin (m\pi x/a)\, dx \right) \left(\int_0^b y(y-b) \sin (n\pi y/b)\, dy \right)$$

$$= \frac{16a^2 b^2 [1-(-1)^m][1-(-1)^n]}{(mn\pi^2)^3}.$$

The general solution (4) will reduce to the form (6) on $z = c$ if

$$C_{mn} = P_{mn} \sinh (b_{mn}c),$$

which means that

$$\phi(x, y, z) = \sum_{m,n} C_{mn} \operatorname{cosech} (b_{mn}c) \sin (m\pi x/a) \sin (n\pi y/b) \sinh (b_{mn}z).$$

Problem (3)

From the same kind of symmetry considerations as in Problem (1) of this chapter, we can assume a potential of the form

$$\phi(r) = A \log r + B, \tag{1}$$

where r is the distance of a point from the axis, and A and B are constants. On the inner cylinder $r = b$ we can put $\phi = V_2$, hence

$$V_2 = A \log b + B. \tag{2}$$

Similarly on the outer cylinder

$$V_1 = A \log a + B. \tag{3}$$

From equations (1) to (3) we see that

$$A = (V_1 - V_2)/\log (a/b). \tag{4}$$

Now the radial component of the electric intensity is given by

$$E_r = -\frac{d\phi}{dr} = -\frac{A}{r}$$

$$= \frac{V_2 - V_1}{r \log (a/b)} \text{ V/m,}$$

and $E_\theta = E_z = 0$.

9

Problem (4)

Suppose we consider first a point charge $-e$ at the point $\mathbf{r}$ and a point charge $+e$ at the point $\mathbf{r}+\mathbf{dr}$, where $\mathbf{dr}$ is an infinitesimal vector. The force on the charge e is $e\mathbf{E}(\mathbf{r}+\mathbf{dr})$. If we only consider the x component of force, we can write

$$E_x(\mathbf{r}+\mathbf{dr}) = E_x(\mathbf{r})+dx\frac{\partial E_x}{\partial x}+dy\frac{\partial E_x}{\partial y}+dz\frac{\partial E_x}{\partial z}+\text{(terms of second order)}$$

$$= E_x(\mathbf{r})+(\mathbf{dr}.\boldsymbol{\nabla})E_x(\mathbf{r})+\cdots.$$

Similarly for the other components of $\mathbf{E}$. Thus to first order the force on the charge e is

$$e[\mathbf{E}(\mathbf{r})+(\mathbf{dr}.\boldsymbol{\nabla})\mathbf{E}(\mathbf{r})].$$

The force on the charge $-e$ is $-e\mathbf{E}(\mathbf{r})$. Hence the total force on the system is (to first order):

$$e(\mathbf{dr}.\boldsymbol{\nabla})\mathbf{E}(\mathbf{r}).$$

In the limit, as we approach the dipole configuration $e\,\mathbf{dr}\to\mathbf{m}$, the electric moment of the dipole and the higher order terms tend to zero, giving for the total force on the dipole

$$(\mathbf{m}.\boldsymbol{\nabla})\mathbf{E}(\mathbf{r}).$$

Problem (5)

The solution of any problem in electrostatics normally entails the solution of Poisson's equation

$$\nabla^2\phi = -\rho/\epsilon_0,$$

given either
    (i) the potential of each conductor; or
    (ii) the total charge on each conductor.
We shall now show that the solution to such a problem is unique.

Let ϕ_1 and ϕ_2 be two solutions of the problem. Then each of these satisfies Poisson's equation and the boundary conditions (i) and (ii). Let $f=\phi_1-\phi_2$. Then f satisfies Laplace's equation

$$\boldsymbol{\nabla}^2f = 0$$

at all points of space. From vector analysis we know that

$$\boldsymbol{\nabla}.(f\boldsymbol{\nabla}f) = f\boldsymbol{\nabla}^2f+(\boldsymbol{\nabla}f)^2 = (\boldsymbol{\nabla}f)^2.$$

Let S_i be the surface of the ith conductor, and S a large spherical surface whose centre is at the origin and which completely encloses the system of conductors. If we let V be the volume between the surfaces S_i and S, and use the above equation, we obtain

$$\iiint_V (\nabla f)^2 \, d\tau = \iiint_V \nabla \cdot (f\nabla f) \, d\tau$$

$$= \iint_S f\nabla f . d\mathbf{S} + \sum_i \iint_{S_i} f\nabla f . d\mathbf{S}. \tag{a}$$

The last equality was obtained by use of the divergence theorem.

Let us now consider the second surface integral in equation (a). If we are given the potential of each conductor as a boundary condition, then $f=0$ on each conductor and this surface integral is zero. If we are given the total charge on the ith conductor, say Q_i, then since ϕ_1 and ϕ_2 are constant on S_i, so is f, and we can remove it outside the integral sign. From the flux theorem of Gauss we have

$$\iint_{S_i} \mathbf{E}_1 . d\mathbf{S} = -\iint_{S_i} \nabla \phi_1 . d\mathbf{S} = Q_i/\epsilon_0.$$

Similarly

$$-\iint_{S_i} \nabla \phi_2 . d\mathbf{S} = Q_i/\epsilon_0.$$

Hence on subtraction we obtain

$$\iint_{S_i} \nabla f . d\mathbf{S} = 0,$$

and hence the surface integral vanishes in (a), since

$$\iint_S f\nabla f . d\mathbf{S} = f \iint_{S_i} \nabla f . d\mathbf{S} = 0.$$

Hence this surface integral vanishes under both sets of boundary conditions.

Now let us consider the first surface integral in equation (a). If the radius of the spherical surface S (say R) is large, and we are a long way from the conductors, then the system appears as a single point charge and the potential is of order R^{-1}. Since ϕ_1 and ϕ_2 behave as R^{-1} on S, then so does f. The electric intensity behaves as R^{-2} for either solution, and hence so does ∇f. Thus $f\nabla f$ is of order R^{-3} on S.

Now the surface area of S is of order R^2, and hence if we let R tend to infinity the surface integral

$$\int\int_S f\nabla f.\,\mathbf{dS}$$

tends to zero. This result does not depend essentially on the spherical form we took for S. Hence if we consider the limit as R tends to infinity, we must have

$$\int\int\int_V (\nabla f)^2\,d\tau = 0.$$

The integrand is non-negative, and hence must be identically zero. Thus ∇f vanishes in V, which implies that f is constant in V. Since f is certainly zero at infinity, it is zero everywhere. Thus the potentials ϕ_1 and ϕ_2 are everywhere equal, and we have demonstrated the uniqueness of the solution of the problem.

Problem (6)

Suppose we have two concentric spherical conductors, the inner of radius a metres, the outer of radius b metres. Let a charge Q coulombs reside on the inner conductor and a charge $-Q$ coulombs on the outer one. By symmetry, the electric field between the conductors is radial. We easily obtain the electric intensity by applying the flux theorem to a concentric spherical surface lying between the conductors and of radius r. We obtain

$$E(r)4\pi r^2 = Q/\epsilon_0,$$

since the electric intensity only depends on the radial co-ordinate r by symmetry. Hence

$$E(r) = Q/4\pi\epsilon_0 r^2 \text{ V/m.}$$

If we remember that in such a system

$$E(r) = -\frac{d\phi}{dr},$$

we obtain for the electric potential ϕ

$$\phi(r) = Q/4\pi\epsilon_0 r + \text{constant (volts)}.$$

Thus the potential difference between the conductors is

$$\frac{Q}{4\pi\epsilon_0}\left(\frac{1}{a}-\frac{1}{b}\right) \text{ volts.}$$

The ratio of the charge on one of the conductors to the potential difference between the conductors is called the *capacity* of such an arrangement, and in this case we write this as C, where

$$C = \frac{1}{4\pi\epsilon_0}\left(\frac{1}{a}-\frac{1}{b}\right) \text{ farads.}$$

An arrangement of conductors such as we have described is called a *capacitor* or *condenser*, and we can have other systems of different geometries.

CHAPTER 4

Problem (1)

Let us consider a small current loop C carrying a current J. Let **dS** be the elemental plane area bounded by C, and in a sense such that the description of C and the direction of **dS** obey the right-hand rule. Since the current in the element **ds** of C is composed of electrons which move with velocity **v** (say), they will experience a force due to the magnetic field. The amount of charge in the element **ds** is

$$dQ = J\, ds/v. \tag{1}$$

The total force on this charge is

$$dQ(\mathbf{v} \times \mathbf{B}) = J(\mathbf{ds} \times \mathbf{B}). \tag{2}$$

Thus the total moment of the forces acting on the circuit about the origin is

$$\mathbf{G} = J \int_C \mathbf{r} \times (\mathbf{ds} \times \mathbf{B}) = J \int_C [(\mathbf{r}.\mathbf{B})\, \mathbf{ds} - (\mathbf{r}.\mathbf{ds})\mathbf{B}]. \tag{3}$$

If the circuit is small enough so that **B** may be assumed constant over **dS**, we have

$$\mathbf{G} = J \int_C (\mathbf{r}.\mathbf{B})\, \mathbf{ds} - J\mathbf{B} \int_C (\mathbf{r}.\mathbf{ds}).$$

However the last integral is easily shown to be zero, since

$$\int_C \mathbf{r}.\mathbf{ds} = \int\int_S (\nabla \times \mathbf{r}).\mathbf{dS} = 0.$$

Hence

$$\mathbf{G} = J \int_C (\mathbf{r}.\mathbf{B})\, d\mathbf{s}.$$

However for any scalar function ϕ we have

$$\int_C \phi\, d\mathbf{s} = \int\int_S d\mathbf{S} \times \nabla\phi.$$

If we put $\phi = (\mathbf{r}.\mathbf{B})$, then $\nabla\phi = \mathbf{B}$, and

$$\mathbf{G} = J \int\int_S d\mathbf{S} \times \mathbf{B} = \mathbf{m} \times \mathbf{B},$$

where $\mathbf{m} = J\, d\mathbf{S}$ is the magnetic moment of the current loop.

Problem (2)

Let the angle between the vectors $\mathbf{m}$ and $\mathbf{B}$ be θ. We consider only small displacements from the equilibrium position so that $\theta \ll 1$. The couple acting on the dipole (see the previous problem) is $\mathbf{m} \times \mathbf{B}$, and this has magnitude $mB \sin \theta$, tending to restore the dipole to the equilibrium position. Equating rate of change of angular momentum of the dipole to the couple produced by the field, we obtain the equation of motion

$$I\ddot{\theta} + mB \sin \theta = 0.$$

Since $\theta \ll 1$, $\sin \theta \approx \theta$ and we can write this equation in the form for simple harmonic motion:

$$\ddot{\theta} + \omega^2\theta = 0, \qquad \omega^2 = mB/I.$$

From this equation we see that the period of the motion is $2\pi/\omega$.

Problem (3)

The field due to a dipole of moment $\mathbf{m}$ at the point $\mathbf{r}$ is

$$\mathbf{B}(\mathbf{r}') = \frac{\mu_0}{4\pi} \left\{ \frac{-\mathbf{m}}{|\mathbf{r}-\mathbf{r}'|^3} + \frac{3[\mathbf{m}.(\mathbf{r}-\mathbf{r}')](\mathbf{r}-\mathbf{r}')}{|\mathbf{r}-\mathbf{r}'|^5} \right\}.$$

The couple $\mathbf{M}$ exerted by this field on a dipole of moment $\mathbf{m}'$ at the point $\mathbf{r}'$ is $\mathbf{m}' \times \mathbf{B}(\mathbf{r}')$, or

$$\mathbf{M} = \frac{\mu_0}{4\pi} \left\{ \frac{\mathbf{m} \times \mathbf{m}'}{|\mathbf{r}-\mathbf{r}'|^3} + \frac{3[\mathbf{m}.(\mathbf{r}-\mathbf{r}')][\mathbf{m}' \times (\mathbf{r}-\mathbf{r}')]}{|\mathbf{r}-\mathbf{r}'|^5} \right\}.$$

The force exerted on $\mathbf{m}'$ by $\mathbf{m}$ is $(\mathbf{m}'.\boldsymbol{\nabla}')\mathbf{B}(\mathbf{r}')$, where the gradient operator acts on the position vector $\mathbf{r}'$, i.e.

$$\mathbf{F} = \frac{\mu_0}{4\pi}(\mathbf{m}'.\boldsymbol{\nabla}')\left\{\frac{-\mathbf{m}}{|\mathbf{r}-\mathbf{r}'|^3} + \frac{3[\mathbf{m}.(\mathbf{r}-\mathbf{r}')](\mathbf{r}-\mathbf{r}')}{|\mathbf{r}-\mathbf{r}'|^5}\right\}.$$

This result only changes in sign if we interchange $\mathbf{m}$ and $\mathbf{m}'$, which shows the equality of the action and reaction as desired.

Problem (4)

Let the two needles have magnetic moments $\mathbf{m}$ and $\mathbf{m}'$ which lie along their lengths and be suspended at a point P. The two needles will rotate until the direction of the resultant magnetic moment of the system at P lies along the direction of the magnetic field $\mathbf{B}$, since in this position the resultant couple on the system is zero (see Fig. 11).

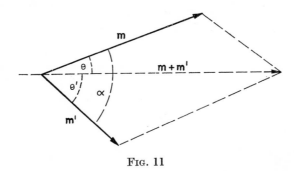

Fig. 11

If we then apply the sine rule to the system in this position we see that

$$\frac{\sin\theta}{m'} = \frac{\sin\theta'}{m} = \frac{\sin\alpha}{(m^2+m'^2+2mm'\cos\alpha)^{1/2}}.$$

Problem (5)

In the equilibrium position the free magnet will lie along a line of force due to the fixed magnet. Since the lines of force due to the fixed magnet, assumed to behave as a dipole, lie in radial planes through its axis, the two magnets will be coplanar. The couple acting on $\mathbf{m}'$ will be

$$\mathbf{G} = \mathbf{m}' \times \left\{-\frac{\mathbf{m}}{r^3} + \frac{3(\mathbf{m}.\mathbf{r})\mathbf{r}}{r^5}\right\}.$$

where **r** is the vector from the fixed magnet to the free one. Hence the magnitude of this couple is

$$G = \frac{mm'}{r^3} (\sin \alpha + 3 \cos \theta \sin \theta')$$

$$= \frac{mm'}{r^3} (\sin \theta \cos \theta' + 2 \cos \theta \sin \theta'),$$

which is zero if $\tan \theta' = -\frac{1}{2} \tan \theta$, as required (see Fig. 12).

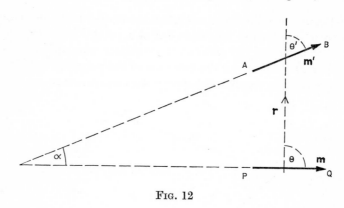

Fig. 12

CHAPTER 5

Problem (1)

For the electric intensity **E** we have

$$\mathbf{E} = \mathbf{E}_0 \sin (pt - \mathbf{k}.\mathbf{r}). \qquad (1)$$

But

$$\nabla \times \mathbf{E} = -\frac{\partial \mathbf{B}}{\partial t}, \qquad (2)$$

and

$$\nabla \times \mathbf{E} = -\mathbf{k} \times \mathbf{E}. \qquad (3)$$

A solution of equation (2) is thus

$$\mathbf{B} = (\mathbf{k} \times \mathbf{E})/kc,$$

since $p/k = c$.

Problem (2)

We have

$$\mathbf{E} = (0, A \cos (pt - kx), A \sin (pt - kx)).$$

Thus

$$\mathbf{\nabla} \times \mathbf{E} = (0, kA \cos (pt - kx), -kA \sin (pt - kx))$$
$$= -\frac{\partial \mathbf{B}}{\partial t}.$$

Hence a solution is, again using $p/k = c$,

$$\mathbf{B} = c^{-1} (0, -A \sin (pt - kx), A \cos (pt - kx)).$$

Note that the end-point of the magnetic vector has a locus similar to that of the electric vector.

Problem (3)

We have

$$\nabla^2 \phi - \frac{1}{c^2} \frac{\partial^2 \phi}{\partial t^2} = 0. \tag{1}$$

Put

$$\phi(x, y, z, t) = \phi_0 \exp \{i[\omega t + kS(x, y, z)]\},$$

with $k = \omega/c$. Then we have

$$\frac{\partial^2 \phi}{\partial t^2} = -\omega^2 \phi, \tag{2}$$

and

$$\frac{\partial \phi}{\partial x} = ik\phi \frac{\partial S}{\partial x},$$
$$\frac{\partial^2 \phi}{\partial x^2} = ik\left(\frac{\partial \phi}{\partial x} \frac{\partial S}{\partial x} + \frac{\partial^2 S}{\partial x^2}\right) = ik\left[\phi \frac{\partial^2 S}{\partial x^2} + ik\phi\left(\frac{\partial S}{\partial x}\right)^2\right], \tag{3}$$

with similar expressions for

$$\frac{\partial^2 \phi}{\partial y^2} \quad \text{and} \quad \frac{\partial^2 \phi}{\partial z^2}.$$

If we now substitute from (3) and (2) into (1), we obtain

$$(\nabla S)^2 - (i/k)\nabla^2 S = 1$$

as required.

CHAPTER 6

Problem (1)

The work done by the electric intensity in driving the current in the wire is EJ per unit length.

The magnetic intensity due to the current in the wire forms circular lines of force in planes normal to the wire and with centres on the wire. The electric intensity $\mathbf{E}$ and the magnetic intensity $\mathbf{B}$ are perpendicular and obey the right-hand rule. Thus the Poynting vector $(\mathbf{E} \times \mathbf{B})/\mu_0$ points into the wire and represents a flow of energy from the field into the wire. The magnitude of the magnetic intensity at distance r from the wire is, as shown in Chapter 4, $\mu_0 J/2\pi r$; hence the magnitude of the Poynting vector is $EJ/2\pi r$. The rate at which energy flows into unit length of wire, supposed of radius R, is thus equal to the magnitude of the Poynting vector times the area of the surface, i.e.

$$(EJ/2\pi R)(2\pi R) = EJ,$$

which we see exactly provides the balance for the work done in driving the current. This was to be expected from conservation of energy.

Problem (2)

The magnetic field due to a dipole $\mathbf{m}$ at the origin is

$$\mathbf{B(r)} = \frac{\mu_0}{4\pi}\left\{-\frac{\mathbf{m}}{r^3}+\frac{3(\mathbf{m.r})\mathbf{r}}{r^5}\right\},$$

which in our particular case becomes

$$\mathbf{B(r)} = \frac{\mu_0}{4\pi}\left\{-\frac{m}{r^3}\mathbf{i}+\frac{3mx}{r^5}\mathbf{r}\right\},$$

where $\mathbf{i}$, $\mathbf{j}$ and $\mathbf{k}$ are unit vectors along the x-, y- and z-axes respectively. Similarly for the electric dipole

$$\mathbf{E(r)} = \frac{1}{4\pi\epsilon_0}\left\{-\frac{M}{r^3}\mathbf{j}+\frac{3My}{r^5}\mathbf{r}\right\}.$$

Since the Poynting vector is proportional to $\mathbf{E} \times \mathbf{B}$, and we can show that this vector has a vanishing divergence, its flux over any closed surface not containing the origin $r = 0$ is zero from the divergence theorem.

Problem (3)

In Chapter 3 we saw that the mutual potential energy of a system of conductors was

$$W = \tfrac{1}{2} \sum_i e_i \phi_i$$

in the notation used in that chapter. For the case we have, we can put

$$e_1 = Q, \qquad e_2 = -Q.$$

This becomes

$$W = \tfrac{1}{2} Q(\phi_1 - \phi_2),$$

where ϕ_1 is the potential of the inner sphere carrying charge Q, and ϕ_2 is the potential of the outer sphere carrying charge $-Q$. We showed in Problem (6) of Chapter 3 that for our system of conductors

$$\phi_1 - \phi_2 = \frac{Q}{4\pi\epsilon_0} \left(\frac{1}{a} - \frac{1}{b} \right).$$

Thus we have

$$W = \frac{Q^2}{8\pi\epsilon_0} \left(\frac{1}{a} - \frac{1}{b} \right).$$

Now let us consider the energy T residing in the electric field between the conductors. We have shown that for this system

$$E(r) = Q/4\pi\epsilon_0 r^2.$$

Hence

$$T = \iiint \tfrac{1}{2}\epsilon_0 E^2 \, d\tau = \frac{Q^2}{8\pi\epsilon_0} \int_a^b \frac{dr}{r^2}$$

$$= \frac{Q^2}{8\pi\epsilon_0} \left(\frac{1}{a} - \frac{1}{b} \right),$$

and we have the desired result.

CHAPTER 7

Problem (1)

Choose rectangular Cartesian co-ordinates with the magnetic field **B** along the positive z-axis, and with the electric field **E** in the (x, z) plane. Then we have

$$\left.\begin{aligned} \mathbf{B} &= (0, 0, B) \\ \mathbf{E} &= (E \sin \theta, 0, E \cos \theta) \end{aligned}\right\} . \qquad (1)$$

The equation of motion of the particle is

$$m\ddot{\mathbf{r}} = e[\mathbf{E} + (\dot{\mathbf{r}} \times \mathbf{B})],\tag{2}$$

which in conjunction with equation (1) gives us

$$m\ddot{x} = eE \sin\theta + eB\dot{y},\tag{3}$$

$$m\ddot{y} = -eB\dot{x},\tag{4}$$

$$m\ddot{z} = eE \cos\theta.\tag{5}$$

From equation (5) we see that the acceleration in the direction of the magnetic field is constant, and that since the particle starts from rest we can imagine the motion to take place in a plane which moves parallel to the z-axis with constant acceleration $eE \cos\theta$.

From equation (4) we obtain, using the initial conditions at $t = 0$,

$$m\dot{y} = eBx.\tag{6}$$

From equations (6) and (3) we get

$$\ddot{x} + (eB/m)^2 x = eE \sin\theta.\tag{7}$$

Put $(eB/m) = \Omega$, the gyrofrequency of the particle. Then

$$\ddot{x} + \Omega^2 x = eE \sin\theta,$$

which has the solution, again using the initial conditions that it starts from rest at the origin at time $t = 0$,

$$x = (eE \sin\theta/\Omega^2)[1 - \cos(\Omega t)].\tag{8}$$

From equations (8) and (6) we obtain on integration

$$y = -(eE \sin\theta/\Omega^2)[\Omega t - \sin(\Omega t)].\tag{9}$$

Equations (8) and (9) show that the equation of the motion in the accelerating plane is a cycloid.

Problem (2)

Take cylindrical polar co-ordinates with axis on the wire and positive z-axis in the direction of the current flow. The magnetic field due to the current in the wire is

$$\mathbf{B} = (0, \mu_0 J/2\pi r, 0).\tag{1}$$

The equation of motion of the charged particle is

$$m\ddot{\mathbf{r}} = e(\dot{\mathbf{r}} \times \mathbf{B}) = e\frac{\mu_0}{2\pi}\left(\frac{-J\dot{z}}{r}, 0, \frac{J\dot{r}}{r}\right). \qquad (2)$$

Hence we have

$$\ddot{r} = -\dot{z}/qr, \qquad (3)$$

$$\ddot{z} = \dot{r}/qr, \qquad (4)$$

where $q = 2\pi m/e\mu_0 J$.

From equation (3) we see that

$$\dot{z} = -qr\ddot{r},$$

which implies that

$$\ddot{z} = -q\frac{d}{dt}(r\ddot{r}). \qquad (5)$$

From equations (5) and (4) we obtain

$$\frac{\dot{r}}{qr} = -q\frac{d}{dt}(r\ddot{r}),$$

which integrates to give

$$(\log r)/q = -qr\ddot{r} + \text{constant.} \qquad (6)$$

At $r = D$ we have $\dot{z} = 0$ and thus from equation (3) we have $\ddot{r} = 0$. Thus we can evaluate the constant in equation (6) to give

$$\frac{\log(r/D)}{r} = -q^2\ddot{r}.$$

This integrates to give

$$\tfrac{1}{2}[\log(r/D)]^2 = -\tfrac{1}{2}q^2\dot{r}^2 + \text{constant.} \qquad (7)$$

At $r = D$ we have $\dot{r} = V$, and thus we obtain from equation (7):

$$\tfrac{1}{2}[\log(r/D)]^2 = \tfrac{1}{2}q^2(V^2 - \dot{r}^2). \qquad (8)$$

From this equation we see that $\dot{r} = 0$ when

$$r = De^{\pm qV} = De^{\pm p},$$

which must be the limiting distances of the particle from the wire.

INDEX